NANOIMPRINT BIOSENSORS

NANOIMPRINT BIOSENSORS

The Fusion of Nanofabrication, Nanophotonics, and Nanobiology

Takeo Nishikawa
Satoshi Fujita

PAN STANFORD PUBLISHING

Published by

Pan Stanford Publishing Pte. Ltd.
Penthouse Level, Suntec Tower 3
8 Temasek Boulevard
Singapore 038988

Email: editorial@panstanford.com
Web: www.panstanford.com

British Library Cataloguing-in-Publication Data
A catalogue record for this book is available from the British Library.

Nanoimprint Biosensors: The Fusion of Nanofabrication, Nanophotonics, and Nanobiology

ISBN 978-981-4316-67-5 (Hardcover)
ISBN 978-981-4463-42-3 (eBook)

Printed in the USA

Contents

Preface

In recent years, the correlation between the health condition of human beings and biological molecules such as nucleic acid and protein, which are major components of the human body, is becoming clearer due to advances in medical technology. Some of these biological molecules called tumor markers are being used as diagnostic indicators in medical tests. In the near future, by detecting the kind, amount, and condition of biological molecules with higher sensitivity and higher accuracy, it would be possible during preventive medical care to remedy and remove a disease before its emergence. Currently, such detection in biological molecules is difficult and costly, and implementing biosensors for the purpose will be greatly helpful in controlling health conditions, reducing medical costs significantly, and improving quality of life of human beings. However, for detecting target molecules with diameter as small as a few to a few tens of nanometers and in low concentration, biosensors with extremely high sensitivity are required. Furthermore, the sensing system should also be compact and popular priced to make its usage widespread.

We have been working to achieve these technically challenging issues by utilizing "nanotechnology" that has been keenly focused upon in this decade. In the conventional biosensing system, the detecting system has a dimension larger than millimeter scale while that of the target molecule is in nanometer scale. Therefore, many additional functions and elements are necessary to compensate the error factors derived from the difference in the detection scale in this system. This results in an increase in the apparatus size and consequently high cost. In our method, the target molecules are detected by a sensor device in which nanopatterns are formed by nanofabrication technology. The nanopatterns are made to interfere with light incident on a nanoscale on the basis of nanophotonics principle and thus ambient minute changes are detected. In this method, since the dimensions of the detection system and the target molecules are in the same order, superfluous noise factors can be eliminated and a high signal-to-noise ratio can be realized. The size of the biosensor system can be as small as a palm in this technology. To achieve high sensitivity, we have also developed a biological probe layer on the sensor surface and it plays the most important part in the procedure which is to capture the target molecules in a specific way. In this biosensor, many different kinds of technologies are included, which have been presented in the book. We feel that fusion of the state-of-the-art technologies with nanofabrication, nanophotonics, and nanobiology fields will be the key to realize new biosensors to satisfy detection demands of the next decade.

In Chapter 1, nanofabrication technologies, especially the nanoimprint technology that can realize mass production of the devices with nanoscale patterns on it, are introduced. Although these technologies have been applied to various applications, such as recording medium, much smaller-scale patterns are necessary to use them for the detection of biological molecules. In addition, a high process throughput is also required to achieve low-cost sensor devices. Examples of basic technologies and applications are also introduced in this chapter.

In Chapter 2, behavior of light in nanoscale is described. In regions smaller than its wavelength (several hundred nanometers), light exhibits "wave" nature. As one of its interesting phenomena, it can resonate with free electrons in a metal. This phenomenon is called "surface plasmon resonance" and is applied to detect the change in ambient resonant condition. However, the resonant interaction of light with free electrons, which is confined in the nanoscale metal structure, has not been analyzed in detailed so far and is being keenly focused upon in recent years.

In Chapter 3, recent trends in biotechnology to develop a surface membrane that can be used on a sensor to capture target biomolecules is introduced. To realize a highly sensitive biosensor, the construction of a probe layer on the basis of an oligo(ethylene glycol)-mixed self-assembled monolayer is very significant. Our developments to achieve a probe layer with high sensitivity and high specificity are also included in this chapter.

In Chapter 4, we introduce the nanoimprint biosensor which we have proposed. As mentioned above, detailed analysis of resonance between light and the nanopatterned metal has not been conducted yet. In this chapter, results of the design, fabrication, and evaluation of the sensor device using nanoimprint technology are presented along with a description of the optimization of nanopatterns for the nanoimprint biosensor.

In Chapter 5, proto-type systems for the nanoimprint biosensors are presented. We have constructed three different types of the proto-models. The first one is for laboratory use, the second for on-site use, and the third for portable use. We have developed a palm-sized biosensor system especially for portable use.

In Chapter 6, applications of nanoimprint biosensors are presented. Detection results of biological molecules using a fabricated sensor are reported. The chapter discusses in detail the use of gold nanoparticles and magnetic beads with nanoimprint biosensors to achieve higher sensitivity, although such systems are still being developed. Besides their use in medical care, application concepts of nanoimprint biosensors in other fields are also introduced in this chapter.

The technologies and concepts of nanoimprint biosensors to detect small molecules can be applied to a wide range of fields such as environmental sensing, food analysis, security check, as well as medical care and healthcare in the future. We hope that this book proves helpful for researchers working on this concept and can thus contribute to improve human life.

Takeo Nishikawa
Satoshi Fujita

Acknowledgement

First of all, we would like to express our gratitude to the publisher of this book, Mr. Stanford Chong and the editorial team of Pan Stanford Publishing for their long-term patience and kind support to complete this book.

We are deeply grateful to Prof. T. Yanagida whose comments and suggestions were of inestimable value for our study. Special thanks also go to Prof. S. Norioka who has kindly helped to initiate our research. We want to thank Dr. T. Wazawa for his valuable technical advice in the biological field. We are indebted to Drs. H. Sezaki and Y. Katsura for their great support in this research.

Special thanks also go to Dr. S. Aoyama for impassioned management of this research project. We appreciate the feedback offered by Dr. Y. Okuno especially on surface plasmon resonance technology. We have greatly benefited from Mr. D. Uchida especially on the application research for our technology.

We are deeply grateful to Mr. T. Matsushita for his leadership in starting and continuing this study. We would particularly like to thank Ms. Y. Haneda for her tremendous help in this research and kind support for establishing our research environment. We would also like to thank Mr. H. Yamashita for his long-term work and contribution on this research. We received generous support from Mr. S. Kamiyama for realizing the nanopatterns in a very small scale. We owe a very important debt to Mr. R. Hasui for his enormous success especially in the construction of proto-models of nanoimprint biosensors. We would like to express our gratitude to Ms. M. Taniguchi for her great work on DNA evaluation using nanoimprint biosensors. We want to thank Ms. R. Masuda for her aggressive study in biological process. We have had the support and encouragement of Ms. M. Kunugiza in various experiments. We want to thank Mr. J. Kishimoto for his great work on waveguide SPR biosensor. Ms. M. Nakamura gave us technical support on fabrication of devices. We could spend enjoyable time with Mr. H. Iwasaka discussing future concepts of biosensor systems. We respect Mr. K. Yamada and Mr. N. Ohgami for their patient works on the development of biosensors.

We would also like to express gratitude to our family for their moral support and warm encouragement. Finally, we would like to thank CREST, Japan, for a grant that made this study possible.

Chapter 1

Nanofabrication and Nanoimprint Technology

The progress of the nanofabrication technology is remarkable especially in this decade. In this chapter, recent nanofabrication and nanoimprint technology is introduced as the basis of nanoimprint biosensor.

1.1 Nanofabrication Technology

The invention of the Schottky diode has changed the world of electric and electronic products and devices as it proved that the miniature tool can be alternative to the traditional vacuum bulb. It, furthermore, proved that the small-sized and integrated components can control the electrons at much higher frequency with much lower power consumption. Since 1970s, LSI (large-scale integration), in which more than one thousand functional elements are integrated in one device, has been developed keenly.

One of the most important characteristics of LSI is that the smaller the device is, the more merits are provided.

When the size of the device becomes k times smaller,

- The planar dimension of the device can be $1/k^2$.
- The delay time to control the electrons can be $1/k$.
- The power consumption can be $1/k^2$.

Nanoimprint Biosensors: The Fusion of Nanofabrication, Nanophotonics, and Nanobiology
Takeo Nishikawa and Satoshi Fujita

ISBN 978-981-4316-67-5 (Hardcover), 978-981-4463-42-3 (eBook)
www.panstanford.com

The products consisting of these devices can be not only smaller but also more inexpensive as the number of fabricated device can be larger on one wafer. And the more rapid calculation can be realized with smaller devices. Though one challenge exists in the heat evolution in the device, many merits are afforded by downsizing the size of device.

So, the history of the LSI development can be said as the history of downsizing. Gordon E. Moore, who is an Intel co-founder, has predicted that the number of the transistor on an integrated circuit increases exponentially, doubling approximately every two years (Moore, 1965). In the LSI field, the microfabrication technology has been rapidly progressed in line with Moore's law. The advanced technologies established in this area were adapted to the other application fields and they provided new discoveries and advances there. In this section, the typical fabrication technologies for the micro- and nanodevices and their trends are introduced.

1.1.1 Photolithography

The most common technique to produce LSI and various semiconductor products at present is photolithography, which uses light to curve the circuit patterns. In this method, the mask substrate which is also called as "reticle" is prepared. Generally, the pattern layout is prepared by chromium layer on the quartz glass substrate. Figure 1.1 shows its principle. The photoresist is spun coated over a silicon wafer in advance. And then the mask pattern is projected onto the photoresist at a miniaturized scale. The mask pattern is transferred onto the photoresist by the exposure of light and developing procedure. To improve the exposure resolution, various research and developments have been addressed in the photoresist and the exposing equipment which is called as "stepper." Since the pattern size cannot be smaller than the diffraction limit of light inherently, the wavelength of the light source has become shorter and shorter in these decades.

Next to g-line (wavelength, 436 nm) and i-line (wavelength, 365 nm) light sources of a high pressure mercury vapor lamp, KrF (wavelength, 248 nm) and ArF (wavelength, 193 nm) light sources by an exima laser are developed and came into practical use now. To realize higher resolution, immersion lithography techniques, which use liquid solution (purified water, oil, etc.) in the exposure chamber and shorten the wavelength of the light source artificially, have been developed. Though these methods are thought to be the most promising way to improve the pattern resolution, other technologies must be added to break through the pattern resolution of sub-10 nm.

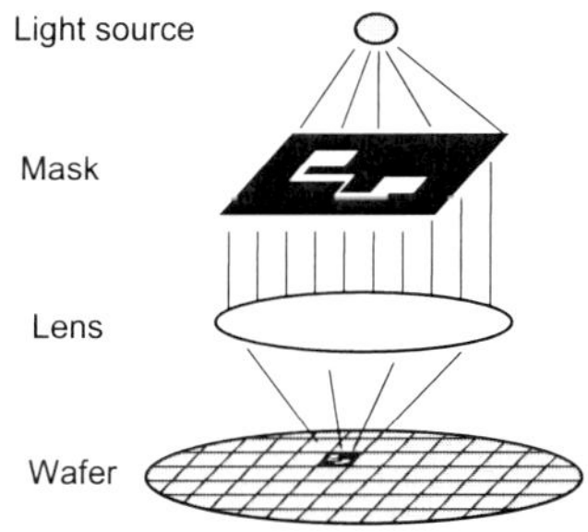

Figure 1.1 Schematic of photolithography process.

To realize sub-10 nm pattern resolution, extreme ultraviolet (EUV) light source whose wavelength is 13.5 nm is being prepared. In this region of wavelength, light cannot be controlled with the refractional optical devices such as lenses. So, the reflection optical control system needs to be constructed. This system must be much more complex than the conventional system (Fig. 1.2). This results in the extraordinarily high pricing of the stepper and this is expected to be a barrier that must be solved to familiarize this method.

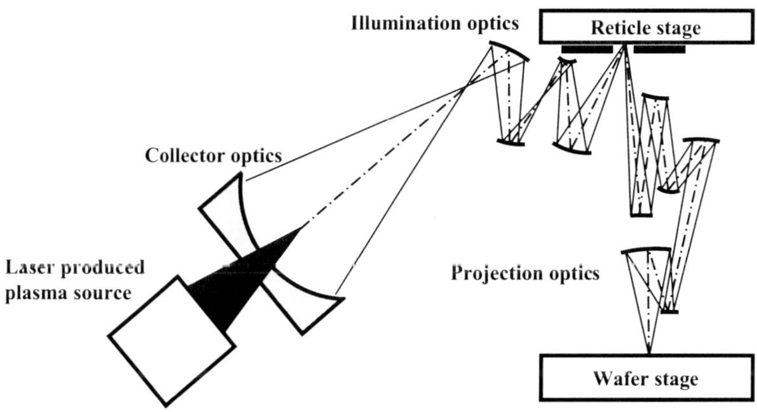

Figure 1.2 EUV exposure system image.

1.1.2 X-Ray Lithography

X-ray lithography which uses X-ray whose wavelength is much smaller than EUV as a light source was proposed in 1972 (Spears and Smith, 1972). Its principle is, basically, same as the photolithography. The X-ray generated by SR-ring goes through the patterned mask and is focused on the resist surface prepared on the wafer as shown in Fig. 1.3. As the wavelength of X-ray is smaller than 1 nm, very high pattern resolution can be theoretically realized. It is, however, pointed out that some severe challenges exist for the preparation of the pattern mask. As the masking material to absorb the X-ray, heavy metal thin layer such as tantalum or tungsten is used generally. To support this masking material, thin membrane (thickness, 0.2–2.0 μm) made of SiN or SiC is prepared under the mask layer as X-ray cannot transmit through the supporting materials such as quartz used in the conventional photolithography.

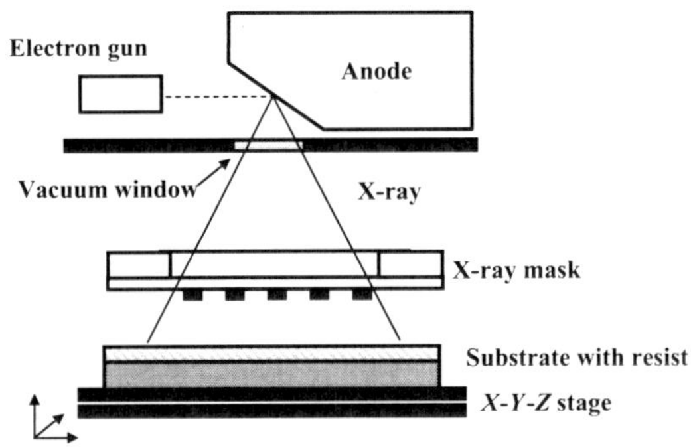

Figure 1.3 X-ray lithography system.

This thin membrane is easily affected by an internal residual stress and results in the variation of the pattern position. As a result, the present accuracy of the mask pattern for X-ray lithography is still several tens of nanometers. Although new mask fabrication techniques using the diamond thin layer as the supporting membrane is proposed and developed, the pattern errors on the mask is still challenging in X-ray lithography.

1.1.3 Electron Beam Lithography

Electron beam (EB) lithography is a technique to write the patterns on the e-beam resist by the EB supplied from the electron gun or electron source. Figure 1.4 shows its schematic image. This method has been used to fabricate the masks for the photolithography. In contrast with the optical photolithography, the wavelength of EB is 0.124–1.24 angstroms when its energy is 10–100 keV. Though the pattern resolution is constrained by the electron scattering in the resist and other aberrations in the system, the pattern size under 10 nm is possible by this method. As the EB is scanned over the resist surface and the patterns are formed in a serial manner, the pattern mask is not necessary. The process throughput is, however, too low to be widely used in the mass production at present. The parallel drawing with multielectron gun is being developed and its realization is expected.

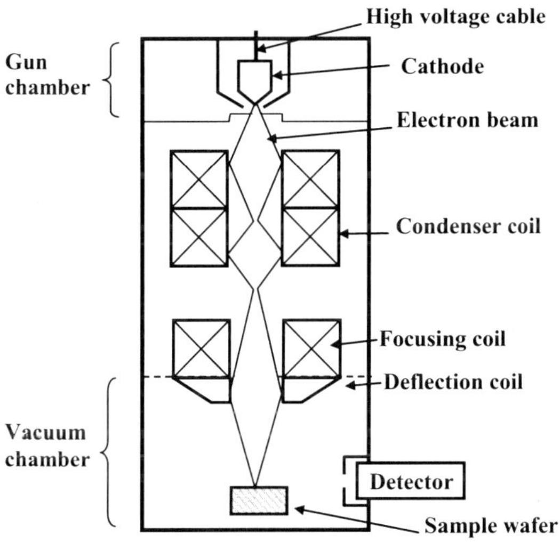

Figure 1.4 Electron beam lithography system.

As mentioned above, the micro- and nanofabrication technologies have been eagerly developed in these decades. And the pattern resolution has been improved into tens of nanometers rapidly. Despite these keen efforts, the technology to realize the patterns of several nanometers in mass production has not been established yet.

In these circumstances, a simple but novel fabrication method based on the different principle was proposed in 1990s. This is the beginning of the nanoimprint world.

1.2 Nanoimprint Technology

Nanoimprint technology was first proposed by Chou et al. in 1995. The principle is that the "mold" which is also called as "template" or "stamper" is pressed onto the surface of polymer and the patterns prepared on the mold surface is transferred onto the polymer surface. The basis of the pattern transfer has been commonly used in various fields. Thousands years ago, the surface of the clay vessel was decorated by transferring the patterns of ropes. Also, printing technique can be said as one of the transferring methods of the patterns from the print plate to the papers. In recent years, the commonplace objects are made from the injection molding which uses three-dimensional mold to copy the plastic or metal products with high process throughput and low cost. Compact disk (CD) and digital video disk (DVD) and Blu-ray disc whose pattern sizes are sub-micrometers are also produced by this method.

Chou has demonstrated that this technique can be applied to fabricate the nanoscale features and that a very high resolution can be realized. Furthermore, it can be used as the pattern lithography by integrating it with the other semiconductor process such as reactive ion etching (RIE). In this section, some major methods of the nanoimprinting are introduced including its history and recent trend of developments.

1.2.1 Thermal Nanoimprint

The principle of thermal imprint has been adopted in the fabrication of the optical storage devices such as the CDs and digital versatile disks whose pattern dimensions are several hundred nanometers. However, the pattern transfer under 100 nm has been never reported before 1995. Professor Chou has discovered that sub-25 nm structures with high aspect ratios can be transferred to the polymer surface. And the new lithography process based on this imprint method was proposed and verified in it. This technology has a potential of a great impact because high mass production of nanodevices can be achieved without highly expensive equipments.

Figure 1.5 shows the fabrication procedure of the nanoimprint technology. The main process flow is as follows:

(i) The thermoplastic polymer (PMMA; polymethyl methacrylate) was applied on to the silicon substrate.
(ii) The mold is heated up to 200 degree Celsius and pressed on to the polymer surface with a high pressure.
(iii) PMMA is cooled down under the glass transition temperature (105 degree Celsius) and the mold is peeled off from the PMMA surface.
(iv) The residue PMMA on the compressed areas is removed by the anisotropic etching.

After this process, the metal nanostructures for the electric circuitry can be prepared by the "lift-off" of the metal layer from this PMMA patterns for electric circuit usages. This fabrication process of the nanosized metal pattern is called "nanoimprint lithography (NIL)."

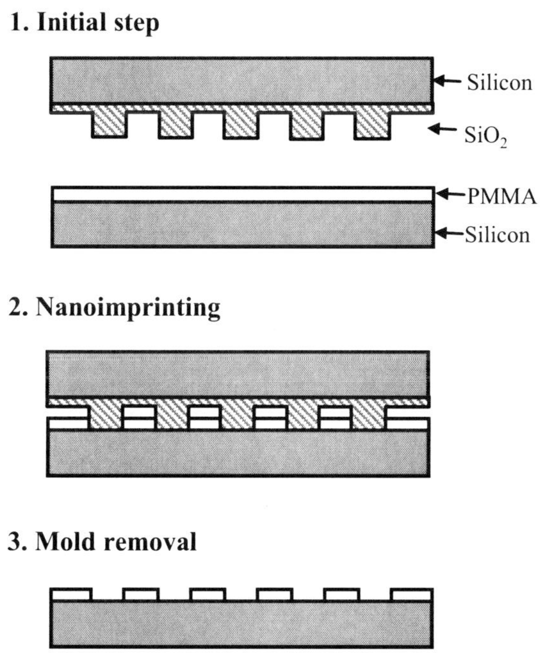

Figure 1.5 Nanoimprint process flow example.

The mold can be made of metals, ceramics, and so on. In this work, the mold of silicon substrate whose surface was covered with the thick silicon dioxide layer was used for the demonstration. The template patterns were prepared by EB lithography and RIE of the silicon dioxide layer. The feature sizes of the fabricated patterns were 25 nm and the depth was 250 nm.

As the thermoplastic polymer, PMMA was spun on a silicon wafer and its thickness resulted in 55 nm. The reasons why they have chosen PMMA for nanoimprint polymer was that it does not adhere well to the silicon dioxide surface and that its shrinkage caused by the temperature changes is not large (about 0.5%). When the polymer is heated above its glass transition temperature, it can behave as a viscous liquid and can pervade into small patterns under the pressure. This realizes the exact pattern transfer from the mold to the polymer surface.

The pressure is also a key parameter to control the pattern transfer performance. Chou has found that the optimum pressure is about 13.1 MPa in his experimental condition. When the pressure is lower than 13.1 MPa, the pattern cannot be fully transferred. When the pressure is too high, the transferred pattern was deformed and nanoscale patterns on the mold could be broken. He could confirm that the dot pattern of 25 nm diameters with 120 nm period could be transferred on that condition.

He also proposed that this process can be used for the lithography techniques. To fabricate the metal nanopatterns, the lift-off technique was adopted after the nanoimprint process. To remove the residue PMMA at the bottoms of the trench regions, oxygen RIE is used. Then 5 nm layer of titanium and 15 nm one of gold were evaporated onto the substrate. As a last step, the remaining PMMA with the covering metal layer was dissolved by being soaked in acetone. The feature size of the metal pattern was about sub-25 nm with 120 nm period as shown in Fig. 1.6 in the first work.

Chou's group, furthermore, presented that sub-10 nm pattern can be also transferred based on this process in 1997 (Chou and Krauss, 1997). This study has proved that the resolution limit of the NIL is decided by the resolution of the mold pattern, not by the imprint process. His first proposal of NIL had a great impact because there had been no other low-cost and high throughput fabrication procedure for fabricating the devices with several tens of nanometer patterns.

As applications of this technology, he has subsequently presented some papers in the late 1990s. For the fabrication of the practical electronic devices, such as memory and transistor, multiple lithography method is necessary. Multilayer resist methods of NIL for the non-flat surface were presented (Chou et al., 1998). In this work, the so-called planarization layer, which is thicker than the height of the pattern step, was prepared as a first step. As the following processes, he proposed and compared five different processes as shown in Fig. 1.7. He also experimentally demonstrated three of them and indicated that the NIL with 10 nm resolution is possible even on the non-flat surfaces (Fig. 1.8).

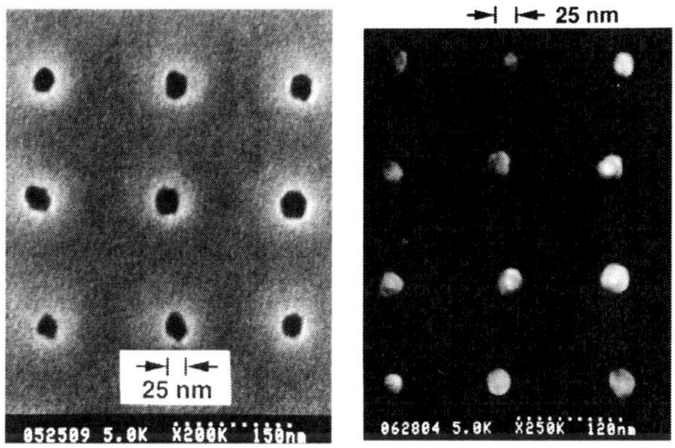

Figure 1.6 SEM micrographs of dot pattern imprinted into PMMA (left) and Ti/Au dot pattern on a silicon substrate (right). Reprinted from Chou and Krauss, 1997, Copyright 1997, with permission from Elsevier.

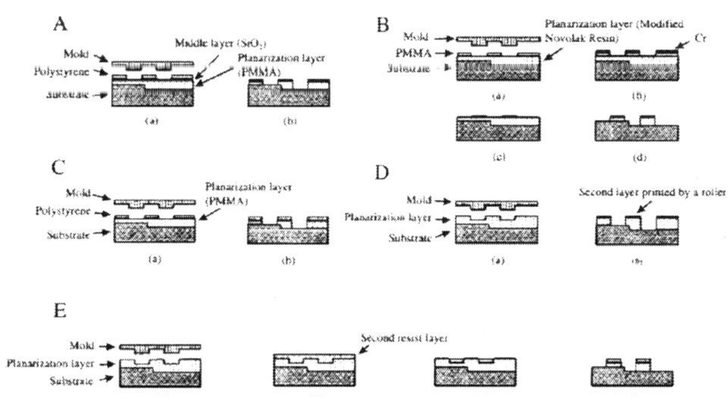

Figure 1.7 Schematic of the multistep nanoimprint processes. Reprinted with permission from Chou et al., 1998. Copyright 1998, American Vacuum Society.

As practical applications, his group has verified that this technology can provide a room temperature single electron memory (Guo et al., 1997) and high density quantized magnetic disks (Wu et al., 1998). These works proved that most of the electronic devices with nanosized structure can be principally produced by NIL and that it has the potential to be alternative to the traditional extraordinarily high-cost lithography method.

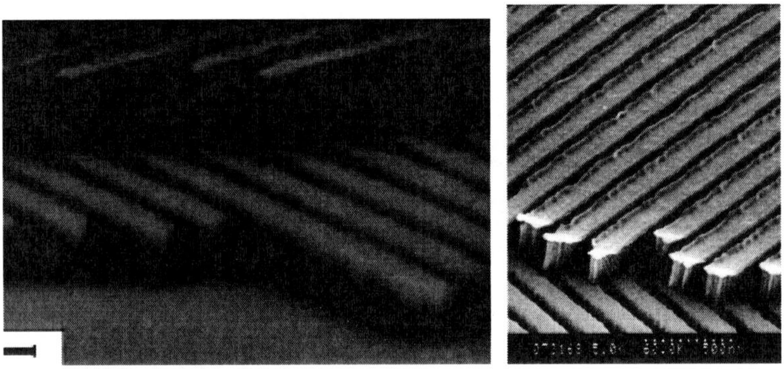

Figure 1.8 Resist grating on a SiO_2 grating in the imprint and lift-off scheme viewed from 70° (left) and 40° (right). Reprinted with permission from Chou et al., 1998. Copyright 1998, American Vacuum Society.

1.2.2 UV-Nanoimprint

While the thermal nanoimprint process has been proved to have the potential to be adopted on the various kinds of electronic device fabrication, some difficulties are pointed out. The challenges in this method are the dead time for the thermal cycle and the alignment errors caused by the heat expansion. The latter can be, especially, a fatal problem for the electronic devices with multiple metal and semiconductor layers. To overcome these problems, a new method which does not use the thermal cycle was demanded.

In 1999, Willson's group proposed another nanoimprint method based on the ultraviolet (UV) curing process (Colburn et al., 1999). The basic protocol for UV nanoimprint is as follows (Fig. 1.9):

(i) The UV curable polymer is dropped onto the substrate.
(ii) A transparent mold is pressed on the polymer with low pressure (less than 0.1 MPa).
(iii) UV light is irradiated through the mold to the polymer.
(iv) After curing the polymer, the mold is released from the polymer surface.

When the metal nanopattern is necessary, the RIE process should be followed subsequently as described in the thermal nanoimprint procedure.

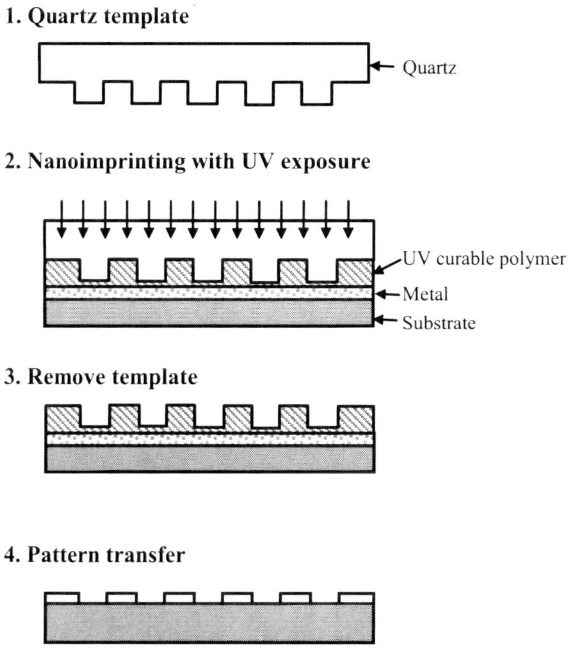

Figure 1.9 Fabrication sequence of UV nanoimprint process.

About the polymer preparation process, another method in which the polymer is spin-coated on the substrate is also proposed (Otto et al., 2000). In this work, the thickness of the residual polymer is theoretically and experimentally studied. A flat surface was achieved by the coating process of polymer in this method. It was, however, found that the fluctuation of the residual polymer thickness and incomplete filling can occur when the geometry of the structures to be imprinted varies strongly across the mold.

This problem is solved by dispensing the polymer with low viscosity on the substrate.

As the mold or substrate must be transparent for the irradiation of the UV light, a quartz mold is generally used. The viscosity of the polymer is so low that the pressure to the mold during the curing process can be very low. Advantageous points compared to the thermal nanoimprint method are as follows:

- The process can be high throughput as the heating and cooling time can be omitted.
- The pattern variation caused through the heat cycle can be suppressed as this process can be executed at room temperature.
- Alignment between the mold and the substrate is possible as the mold is transparent.

As this process just needs the mold control and the UV irradiation, it is called as step and flash imprint lithography (S-FIL). For the use of the lithography, this is considered to be the most suitable method. And after Willson's proposal, this technology has been widely studied and improved including the template lifetime, fabrication throughput, residual material tolerances, and critical dimension control and the equipment.

To realize a high throughput fabrication with a large diameter wafer, step and repeat method is proposed for S-FIL (Colburn et al., 2001). Figure 1.10 explains its process flow. As a first step, the UV curable resin is dispensed on a part of the wafer. Then the patterns are transferred onto the resin surface in this part by using the aforementioned protocol. Then the mold is released and steps to the next field and repeat this cycle up to the end of the wafer. After the wafer is fully imprinted, the etching process is conducted. By this step and repeat method, 200 mm diameter wafer can be imprinted with a small mold with high process throughput.

For the functional electronic devices, multilayer lithography is necessary. This is thought to be the most challenging and important point to be alternative to the conventional lithography methods and keenly addressed in these few years. To solve this problem, Choi et al. have developed alignment mechanisms and methods (Choi et al., 2005) and stage motion control systems (Choi et al., 2008). They reported that the alignment and magnification control error (three standard deviations) can be smaller than 10 nm. Some venture companies have already commercialized the nanoimprint equipments based on S-FIL technology. And they are being applied to the fabrication of light-emitting diode (LED) and high density memory device and so on. It is expected to come into practical use in these few years.

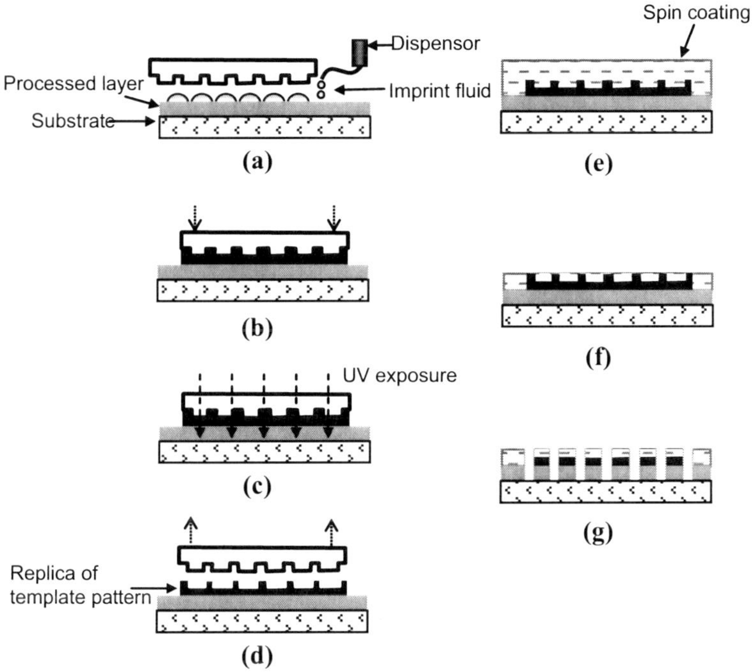

Figure 1.10 Schematic of S-FIL process flow. (a) Dispense the drops of organic containing imprint fluid, (b) superimpose the transparent mold on the imprint fluid, (c) polymerize imprint fluid with UV exposure, (d) separate template from substrate, (e) spin coating silicon-containing material for planarization, (f) halogen etching to expose imprinted features, and (g) oxygen etching to etch organic imprinted pattern and processed layer.

1.2.3 Other Nanoimprint Technologies

Other than the basic thermal and UV NIL techniques, various methods have been developed. Some examples are introduced in this part.

1.2.3.1 Direct metal imprint

In this method, the patterns on the mold are transferred onto the metal surface directly. As an extraordinarily high pressure is applied in this process, hard mold made of SiC or diamond is used in this process. To prepare the mold, the patterns are prepared by the EB and metal patterns are formed by lift-off process. And then, the surface is etched by RIE while the metal patterns work as a high selective etching mask. And they are removed by the following immersion etching process. Masuda et al. has used a SiC mold and demonstrated the nanoscale patterns can be directly replicated on the Al layer (Tamura et al., 1998). The patterns used here was the line and space in which the line width was 40 nm and the pattern period was 250 nm and the dots whose diameter was 40 nm. The applied pressure on the mold was 2.3 GPa. Though a hard mold and high pressure equipment are necessary in this method, the patterns can be directly transferred to the metal substrate without the heat cycle and any UV light irradiations.

Chou et al. devised another direct imprint method that uses laser pulse to melt the surface of the substrate which is called as "laser-assisted direct imprint: LADI" (Chou et al., 2002). In this method, a quartz mold with the nanoscale patterns is brought into contact with the silicon substrate. A single XeCl (308 nm wavelength) laser pulse with 1.6 mJ/cm^2 (20 nm pulse width) melts a thin surface layer of Si. The molten silicon is embossed while the silicon is in the liquid phase in a very short time (less than 250 ns). Then the silicon surface solidifies rapidly. The mold and substrate are separated and the negative profile of the mold patterns was left on the silicon surface. The imprinted feature was 140 nm wide, 110 nm deep and had a 300 nm period. They indicated the LADI process has sub-10 nm resolution from this result. This method has the potential to realize the low-cost and high throughput fabrication process because it does not require any subsequent etching processes.

1.2.3.2 Soft lithography (microcontact printing)

In 1993, G. M. Whitesides proposed another imprinting technique which uses a soft mold made of polydimethylsiloxane (PDMS) (Kumar and Whitesides, 1993). The process flow is shown in Fig. 1.11. They prepared the PDMS mold with microscale patterns by conventional photolithography as the first step. Then they add the alkanethiol ink on the PDMS surface and press the mold onto the thin gold layer on a silicon substrate. Only the protrusion region on the mold contact to the gold surface and the alkanethiol ink in this region is transferred from mold to substrate. The thiol group has the characteristic to bind to the gold covalently and form a self-assembled monolayer whose thickness is about 1.2 nm. When the substrate is immersed in the etchant of gold, the thiol group plays as an etching mask and only the regions where the thiol monolayer is not formed are etched. As the result, the patterns on the mold can be transferred to the gold patterns. The most different point of this method from the thermal and UV nanoimprint is that the soft and elastic mold is used to produce the patterns. This characteristic enables the nanopatterns to be imprinted even on the curved surface. Also, this process can be completed through a room temperature. The biomaterials can be used as "ink" for the biodevices as will be described later.

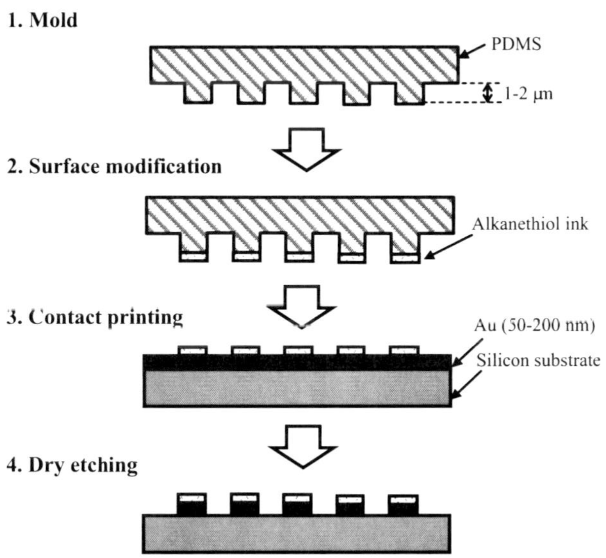

Figure 1.11 Soft lithography process flow.

1.2.3.3 Molecular imprint

While the conventional imprint technology described above can transfer only the shape and sizes of master material, a unique technique called "molecular imprint" has been proposed and developed for transferring chemical function on the master substrate

(Wulff, 1995). The transfer of chemical function can realize the new innovative devices such as artificial enzyme and biomaterials.

Shi et al. (1999) use radiofrequency glow-discharge plasma deposition to form polymeric thin films around proteins coated with disaccharide molecules. It is demonstrated that the disaccharides become covalently attached to the polymer film, creating polysaccharide-like cavities that exhibit highly selective recognition for a variety of template proteins. The artificial surface capable of selectively recognizing proteins can be applied for separations, biosensors, and the development of biomedical materials.

1.3 Key Technologies and Examples of Nanoimprint Technique

Figure 1.12 shows the basic process flow of the nanoimprint method. As the first step, the master patterns need to be fabricated on a substrate. And then, the mold is prepared from the master. The last process is the replication in which many replicas are produced. The key technologies and recent trends including our fabrication examples are introduced in this section.

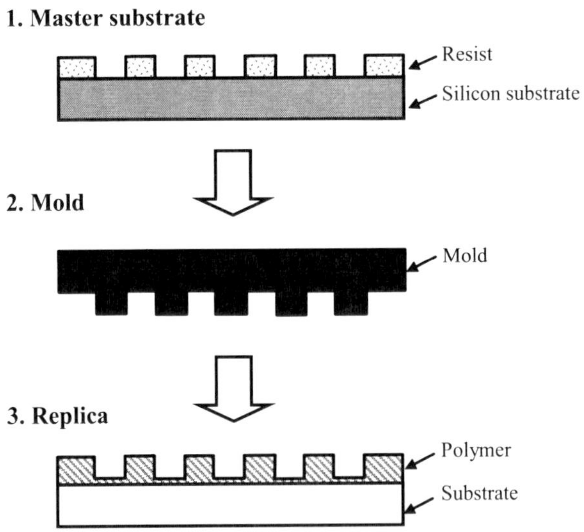

Figure 1.12 Fabrication flow of master, mold, and replica by using nanoimprint technology.

1.3.1 Master Process

Some nanoscale fabrication methods such as photolithography, X-ray lithography, and so on are described in Section 1.1. The most appropriate method will be chosen in consideration with pattern width, pattern area size, process throughput, equipment, and material cost.

As an example, our fabrication result using EB lithography is introduced here. The detail process flow is as follows:

(i) The EB resist is spun on a cleaned silicon wafer.
(ii) The substrate is pre-baked before lithography.
(iii) The requisite patterns are drawn by EB lithography.
(iv) The EB is developed by using chemical agent.
(v) The substrate is post-baked after development.

The thickness of EB resist which determines the height of nanopatterns in this process can be controlled by adjusting the spin-coating conditions (EB resist material, spin speed, ambient temperature). And the baking conditions and EB drawing conditions are also very significant for improving the actual pattern resolutions.

Figure 1.13 shows the SEM images of the fabricated pattern. The pictures are magnified from (a) to (b), (c), (d), (e), and (f), gradually. In (a), a fly is arranged with pattern for the comparison of the size. In this picture, the width of one character is about 500 μm. As shown in (b), the characters consist of smaller characters. The line width of one character in (c) is about 10 μm. These characters, again, consist of much smaller characters as shown in (d) and (e). The line width composing the smallest characters is about as small as 10 nm which is almost same as the size of protein molecules. In this way, the use of EB enables to draw arbitrary nanoscale patterns by preparing the sequence program.

1.3.2 Mold

The mold plays very important role in nanoimprinting technique because the final patterns are mainly determined by the accuracy of the mold. The characteristics requested for the mold is as follows:

- High precision (including pattern errors due to temperature fluctuation).

- High durability.
- Surface detachment characteristic from the replica.

There are two main ways to prepare the mold substrate. One is to use the master substrate as a mold by adding some modification process on it. And the other is to produce another mold substrate from the master substrate.

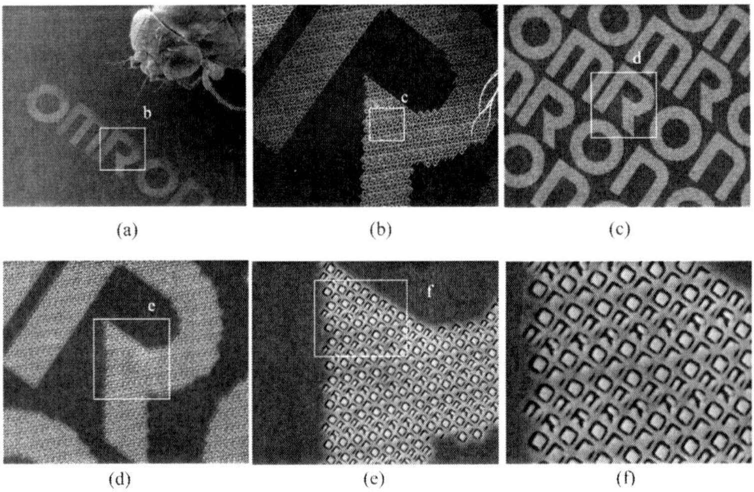

Figure 1.13 SEM images of master substrate example fabricated by EB lithography.

Figure 1.14 shows an example of the typical former processes. As the first step, silicon dioxide layer (thickness, 60 nm) is formed on the surface by heating silicon wafer in high-temperature furnace within an oxygen atmosphere (thermal oxidation). The EB resist such as PMMA is spun on it and developed after drawing the nanoscale patterns by EB process. When using the posi-resist, the exposed area is subject to be dissolved by the developing fluids. Subsequently, a chromium layer is deposited on it. The thickness of chromium layer should be thinner than that of EB resist layer to separate the chromium layer in the bottom of the resist patterns from the chromium layer on top of it. By dissolving the EB resist, only the bottom chromium layer remains (lift-off process). In the next step, the remaining chromium layer works as a mask when removing the silicon dioxide layer by dry etching process. This dry etching is conducted under the atmosphere of CHF_3 (trifluoromethane) which has the characteristic to etch the silicone dioxide faster than the silicon. So, the height of the pattern can be almost equal to the original thickness of the silicon dioxide layer. The mold is completed after removing the chromium layer. In this method, many semiconductor fabrication processes are used and it takes many efforts and long tact time. The completed mold, however, can be used repeatedly in the nanoimprint method. Therefore, it can be sometimes said that the mold with high precision and high durability should be fabricated even if it costs much to prepare it.

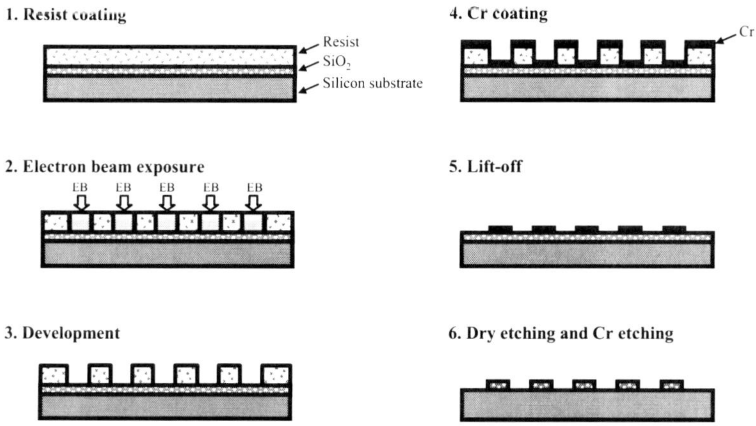

Figure 1.14 Mold fabrication process directly from master substrate.

An example of the latter processes preparing the other mold from the master is to use the electroforming process. Electroforming process is widely used for the metal coating and the replication of surface patterns. Figure 1.15 shows the process flow to fabricate the mold. First, a nickel layer is formed by sputtering to prepare a conducting layer on the master substrate for the electroforming process. The thickness of the sputtered layer is about 200 nm (thoroughly thicker than the height of the nanoscale patterns on the master substrate). The patterns on the master substrate are transferred by the nickel sputtered layer in this method. Then the master substrate is soaked in an electrolytic solution bath. The solution consists of a sulfamic acid in which pH is controlled in about 4.0–4.5, the temperature is 40–45 degree centigrade. The nickel ion (2+) is dissolved in the solution. The sputtered nickel surface on the master substrate is conducted to the cathode through the contact ring arranged around the edge of master wafer. When the voltage is applied between the anode and the cathode electrodes, the nickel ions move toward the cathode (master substrate). And when reaching the cathode, they receive electrons from the electrode and nickel particles are precipitated. The applied voltage is controlled as the electric current can be constant (45 ampere) during the electroforming. This is continued till the thickness of the electroformed layer is about 250 μm, that takes about 1 hour on our condition. This electroformed layer works as a supporting layer for the nickel sputtered layer. Then the mold is released from the master substrate. When the EB resist is used as a master surface, its residue adheres on the mold surface after the separation. By removing them, the mold preparation is completed.

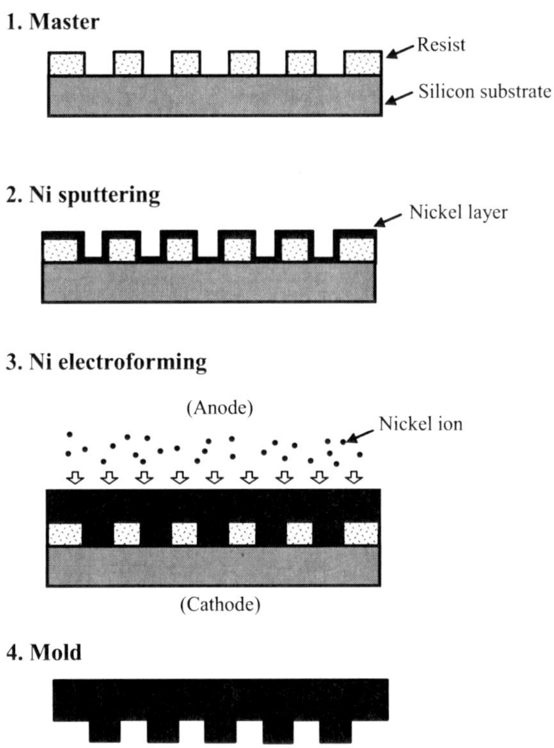

Figure 1.15 Mold fabrication through electroforming process.

The pattern shrinkage caused by the differences of the thermal expansion coefficients between the master substrate (silicon) and mold (nickel) should be taken into consideration in designing the master pattern in advance. Also, various parameters such as temperature, pH, ion concentration, and contamination in solution must be adjusted to control the internal stress of the electroformed layer, which causes the pattern position error and the camber of mold. Figure 1.16 shows the SEM images of the mold surface fabricated from the master substrate introduced in the precious part. We can see the patterns in as small as several tens of nanometers can be transferred onto the mold surface by this procedure.

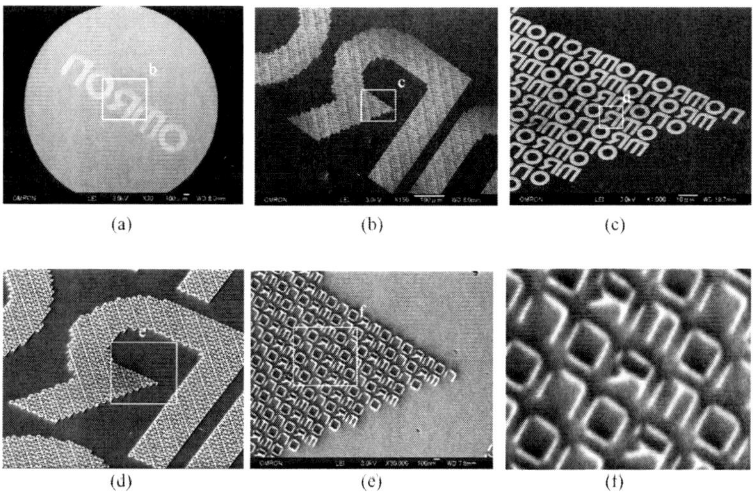

Figure 1.16 SEM images of mold example fabricated by electroforming process.

The advantage of this method is that multiple molds (second generation) can be produced from the mold (first generation) which is prepared from the master. In addition, further multiple molds (third generation) can be produced from the second-generation molds. By repeating the electroforming process, a large number of molds can be produced from only one master in this method as shown in Fig. 1.17. And they are favorable to keep the product's lot quality since the molds fabricated from the same master hardly have the pattern differences.

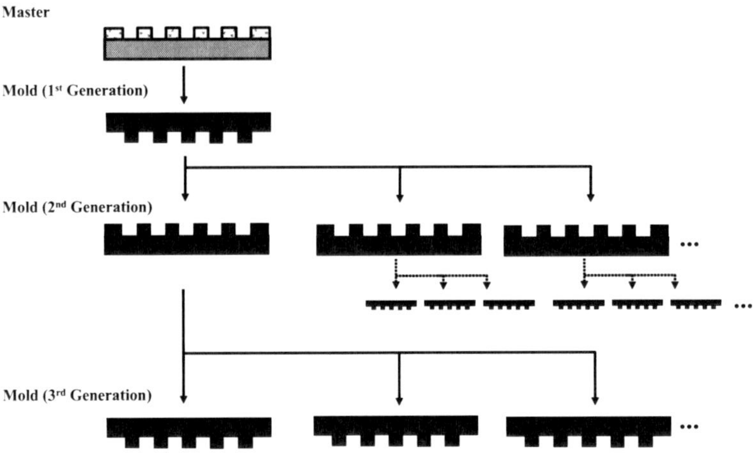

Figure 1.17 Multiple electroforming process for mold fabrication.

1.3.3 Replication and Recent Trends

The basic principle of the replication is to duplicate the reversal topological copies from the mold by using heat or ultraviolet and so on as presented above. The replication process is the final step in the nanoimprint technique that decides the most process throughput capabilities. Many process improvements in this step are, therefore, addressed in these years. And some venture companies have started to commercialize the nanoimprint equipments. The current key issues in the developments are as follows:

(i) Higher process throughput and larger processing area.
(ii) Higher pattern accuracy (reproducibility, alignment accuracy, multilayer overlap accuracy).
(iii) Various pattern imprinting (three-dimensional pattern, high aspect ratio, micro- and nanohybrid patterns) for applications.

Technical approaches for (i) and (ii) are related because the larger the wafer size is, the more challenging to keep high accuracy in the wafer area becomes. And the imprinting methods and systems directly decide these specifications. The recent improvements about them will be introduced in this part. The activities about issue (iii) will be depicted in Section 1.4 including some specific application examples. Realizing the larger processing area can provide a higher process throughput and results in the reduction of process cost for each device. To achieve it, some methods have been proposed such as a way using a large flat-type mold (Heidari et al., 2000), a roller imprint lithography (RIL) which uses rolling cylinder mold (Tan et al., 1998), and a step and stamp imprint lithography (SSIL) technique (Sotomayor et al., 2003) which uses a small mold and align it over the wafer repeatedly.

Flat mold The first way to use a large flat mold is the simplest one. It can realize a seamless pattern with a single step. It, however, requires an extraordinarily high pressing force (about several tons) and the temperature uniformity in a large area especially using a thermal imprinting. Furthermore, the local variation of the flatness over a large mold can lead to the pressure concentration at a local area and results in a distortion of a mold (Deguchi et al., 2002; Lazzarino et al., 2004). Many improvements have been addressed for these problems and recent study reports that they realize a uniform pressure on the substrate by blowing air through the holes in substrate onto the backside of the substrate (Kim et al., 2009). For 18-inch diagonal flat panel displays, they used 390 mm × 490 mm quartz mold and replicated micropatterns in a low vacuum condition to prevent the air bubble from nipping.

Roller mold The second one using a roller cylinder to press the polymer is first proposed by Chou's group. Figure 1.18 shows a schematic of a roller imprint system. A roller cylinder is prepared on the cover side of the system. And its temperature can be controlled by a heater inside it. Underneath it, a flat platform to hold the sample is provided and it possesses ball-ring rails to move back and forth. The cover plate is connected to the base platform via the hinges. The pressure can be controlled by applying weights on the hinges. They have proposed two methods as shown in Fig. 1.19. One is to bend a thin metal film mold around the roller. And the other is to use a flat solid (silicon) mold and to press it from the backside with rotating and pressing by the roller. In both methods, they have achieved the replication with sub-100 nm pattern resolution. They found that few air bubbles remain on the replica as the roller pushes air out. And another good point in this method is that the pressure concentrates in the small area where the roller contacts to the substrate and an extraordinary pressure is not necessary in this method. It can offer a much simpler and higher throughput and low-cost machine. A recent study reports that the RIL can be achieved in a large pattern area (100 mm × 100 mm) by adjusting the pressure and roller moving speed (Youn et al., 2008).

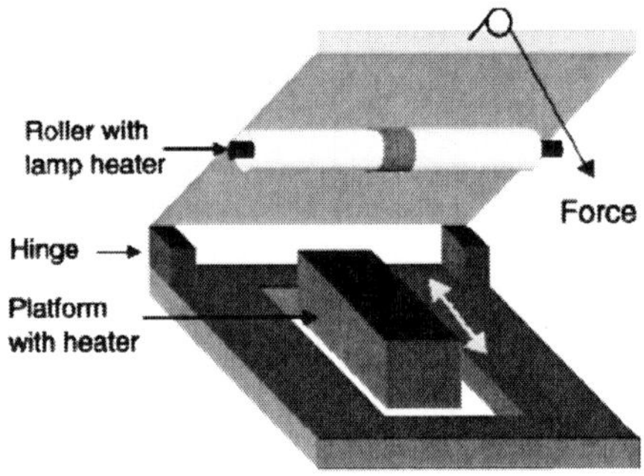

Figure 1.18 Schematic of roller nanoimprint system. Reprinted with permission from Tan et al. 1998. Copyright 1998, American Vacuum Society.

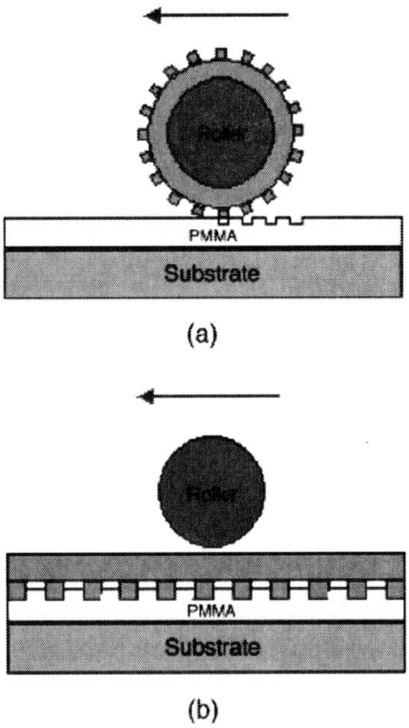

Figure 1.19 Schematics of two ways for roller nanoimprint process. Reprinted with permission from Tan et al. 1998. Copyright 1998, American Vacuum Society.

Step and repeat The third way is to repeat the alignment step and imprinting step over a large wafer by using a small element mold (Colburn et al., 2001). Some venture companies (Molecular Imprints Inc., Nanonex, etc.) have commercialized the nanoimprinting machines equipped with step and repeat mechanism. Figure 1.20 explains its basic process flow with UV imprinting. As the first step, the UV curable resin is dispensed on a part of the wafer. Then the mold is pressed onto the substrate and the patterns are transferred onto the resin surface by UV irradiation in this part. Then the mold is released and steps to the next field and repeat this cycle up to the end of the wafer. As the mold can be small, the mold fabrication can be relatively cost-effective. And the uniform pressure control can be easier than that with a larger mold imprinting. On the other side, the process throughput goes down when the element mold is too small. Lee et al. (2006) have used 5 × 5 inch square mold to improve the process throughput, whereas the typical element mold was 1 × 1 inch square. They demonstrated that 50–80 nm features were successfully transferred on the 8 inch wafer by only four imprint steps by this mold.

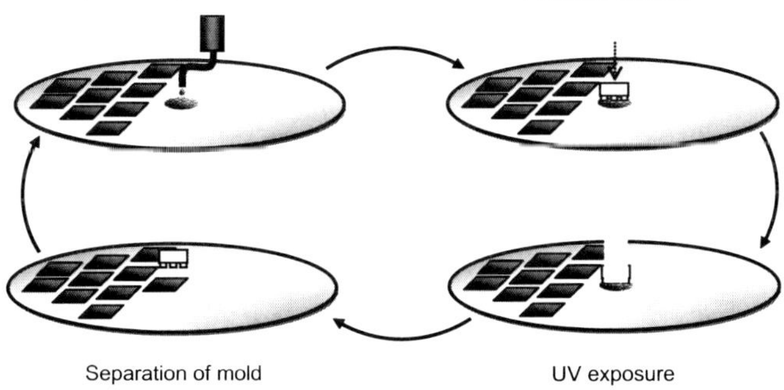

Figure 1.20 Process flow of step and repeat imprint technology.

We have also developed a step and repeat thermal imprinting equipment with Engineering System Corporation (Nagano, Japan) as shown in Fig. 1.21. Its dimension is about 5 (wide) × 4 (depth) × 3 (height) m^3 and its weight is 15 tons. It is located in a clean room to prevent the mold and replicas from being contaminated by the dusts. Its press pressure can be up to 250 tons and the substrate size can be as large as 1 m × 1 m. Three different molds whose sizes are about 200 mm diameter can be set simultaneously on the mold holding chucks and they can be selected according to the pattern designs during the process flow. After setting the polymer coated substrate on a stage, the substrate and mold are heated with the heater and the mold is pressed on the substrate. Figure 1.22 shows an example of an optical device made by this machine. To control light to a specific direction, sub-micropatterns were produced in a large area. This can be used as the backlight panel for the mobile phone. By controlling the light direction correctly by these patterns, the display can be brighter and more beautiful than the conventional randomly patterned backlight substrates.

Figure 1.21 Nanoimprint system for large-scale substrate.

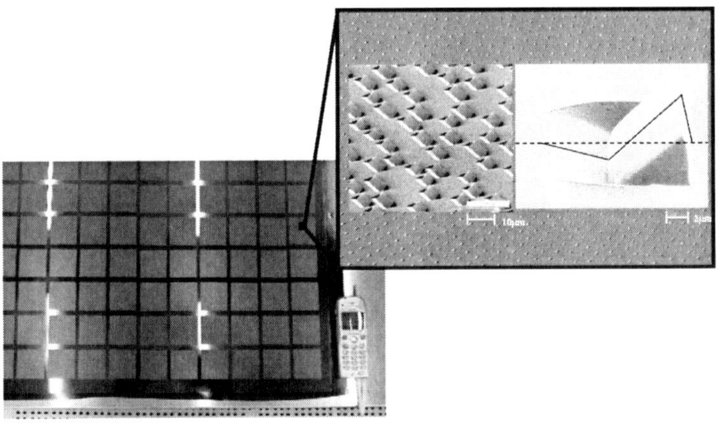

Figure 1.22 Backlight module substrate and the patterns on the substrate fabricated nanoimprint process.

1.4 Applications of Nanoimprint Technology

Since the proposal of nanoimprint technology in 1990s, it has been keenly addressed and developed. And many applications have been studied and developed in various areas as given below.

(i) Electric devices (CMOS, transistor, patterned media, etc.).
(ii) Optical and photonic devices (microlens array, polarizer, photonic crystal, anti-reflection structure, hologram, etc.).
(iii) Biological devices (filter, microchannel, molecular imprint).
(iv) Energy devices (solar cell, fuel cell, LED).

In this section, the examples of these applications and the derivative technologies are introduced.

1.4.1 Electron Devices

As mentioned above, the conventional photolithography is encountering the difficulties for its extraordinary equipment cost. NIL is expected to be an alternative for it, and the demonstrations of the electron devices (CMOS, FET, memory, patterned media, etc.) based on this fabrication technique are reported.

1.4.1.1 Filed emission transistor

Field emission transistor (FET) is one of the key electron devices in the present integrated circuit. FET is a type of transistor which controls the electron current from source to drain by adding a voltage on the gate channel. The smaller the gate size is, the faster is its response. The downsizing of the gate dimension is, therefore, focused on its development.

A nanoscale silicon FET by NIL was first demonstrated in 1997 (Guo et al., 1997). The transistors were prepared on a silicon-on-insulator (SOI) wafer using a three-level lithography process. The first layer which is the active areas of transistor including nanoscale channel features was fabricated by NIL as presented in Fig. 1.23. As the second and third layer, the gates and the final metal contacts were patterned by the conventional photolithography. The metal oxide semiconductor FET (MOSFET) with 100 nm gate width was fabricated by this procedure. They found that this device has almost identical characteristics with the conventional FET fabricated by e-beam lithography.

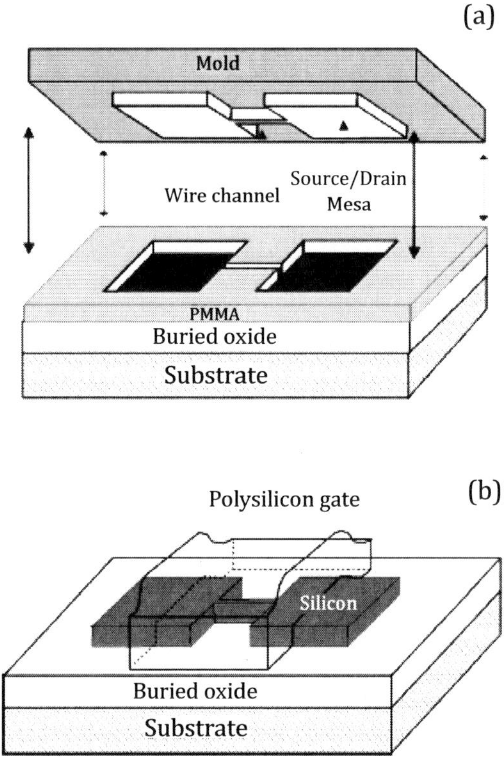

Figure 1.23 Imprint process for nanoscale FET. Reprinted with permission from Guo et al., 1997. Copyright 1997, AIP Publishing LLC.

Additionally, the MOSFET whose all layers were fabricated by NIL over the entire 4 inch wafer was reported in 2003 (Zhang and Chou, 2003). This study addressed the issues of overlay accuracy, large area uniformity, and low defect density which are thought to be the most important to realize the commercialization of the NIL electron devices. To minimize the alignment error caused by the heating and cooling cycle, the mold made of Si which has an identical thermal expansion coefficient for the nanoimprinting on a Si wafer was employed. Furthermore, to avoid any displacement between the mold and wafer, very uniform pressing force over an entire wafer was applied during NIL by home-built nanoimprint machine. As the Si mold and substrate is not transparent, an image processing alignment technique called as "back-side alignment" is used. In this technique, the wafer is placed first and its alignment mark image was captured by microscope camera, followed by inserting the mold and aligning the alignment marks on the back side of the mold with the captured images of alignment marks on the wafer. By these improvements, the FET with 60 nm gate width on 4 inch wafer was fabricated within 500 nm alignment error as shown in Fig. 1.24. This device was demonstrated to work as a transistor properly, that means NIL can be used for the multilayer electron device on a large wafer size.

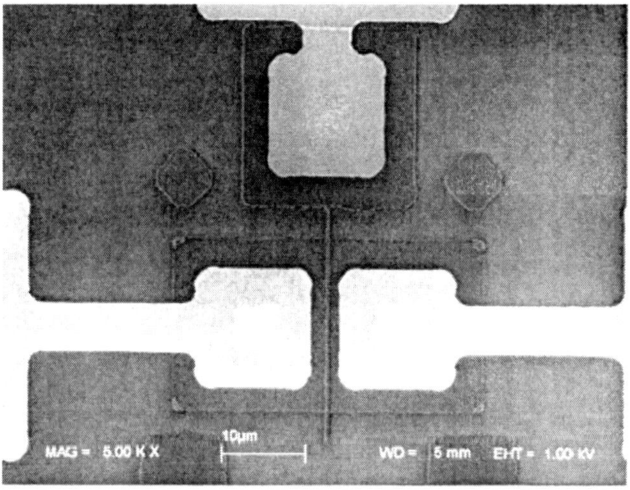

Figure 1.24 SEM image of MOSFET fabricated by nanoimprint lithography. Reprinted with permission from Zhang and Chou, 2003. Copyright 2003, AIP Publishing LLC.

1.4.1.2 Patterned media

Hard disk drive in which the data information is stored by the magnetic polarity of each grain was invented by IBM team in 1954. The first disk storage unit "IBM 350" which realized a real-time and random reading and writing data was produced in 1956. It consists of fifty 24-inch (610 mm) diameter disks and its storage volume was about 4.4 Mbytes. In the last decades, the recording density of the hard disk drives has enormously increased and they widely spread as the common data storage devices for wide variety of applications such as personal computer, car navigation system, TV and video recording system, portable music player, and so on.

The downsizing of its dimension and the increase of the storage volume have been achieved by scaling down the grain size in the media and by increasing the recording density. Though this is a simple and effective approach, it is pointed out that there exists a limitation in this method caused by the superparamagnetic effect, which means that the magnetization cannot be kept in a long term for thermal instability when the grain size is too small.

To overcome this problem, a new recording paradigm of patterned media in which the magnetized cells are separated by a nonmagnetic substance has been proposed as the potential ways for further increasing areal recording density (Ross et al., 1999; Lodder, 2004).

In this method, the discrete islands of magnetic substance corresponding to the recording bits are preformed on the media's surface. The more high recording density to realize, the higher resolution of pattern lithography is demanded. As a first demonstration, EB lithography was used to fabricate 65 Gbit/in^2 patterned media in 1995 (Krauss and Chou, 1995). Though it was verified that the patterned magnetic substance can store the magnetization, this process is too expensive to be used for the commercialization of patterned media as its process throughput is too low. Another fabrication way is to use laser interference lithography (Chou et al., 2002; Grigalinas et al., 2004). This method can provide a relative high process throughput in a large area. The pattern cross-sectional shape must be, however, sinusoidal and this confines the pattern design flexibility and precise fabrication control of pattern sizes. As a promising way, a fabrication procedure using nanoimprint technology was suggested and demonstrated in 1998 as introduced in Section 1.2.1.

Wu et al. (1998) have demonstrated to use the NIL for the patterned media, Fig. 1.25 shows its process flow. First, a thin metal layer for the plating base was deposited on a flat silicon substrate. And the SiO_2 layer which works as a nonmagnetic wall was prepared. After that 200 nm PMMA resist was coated on it. Second, a nanoimprint mold with a period of 190 nm and a height of 180 nm was used to transfer the nanosized features. Then O_2 RIE was conducted to remove the PMMA at the bottom of the patterned holes. Third, 20 nm chromium layer was prepared only on the top of PMMA by evaporation at a glancing angle to provide an additional mask for etching SiO_2. Fourth, CHF_3 RIE etching etched SiO_2 at the bottom of patterned holes and was stopped at the plating base. Fifth, nickel nanopoles were formed by electroplating at the exposed plating base. Finally, chemical mechanical polishing was used to remove the excess nickel to realize a smooth and flat surface. This procedure has achieved 18 Gbit/in^2 (Fig. 1.26) with a perpendicular quantized magnetic disk and 30 Gbit/in^2 with a longitudinal quantized magnetic disk. Though they are sufficiently high density at the published time, the processes such as electroplating and CMP (chemical mechanical polishing) are relatively complex and time consuming for the mass production of magnetic disks. Furthermore, to achieve higher recording density, smaller lithography resolution is necessary (Terris, 2009). When the pattern resolution can be smaller than 10 nm, patterned densities as high as 10 Tb/in^2 can be possible by this method.

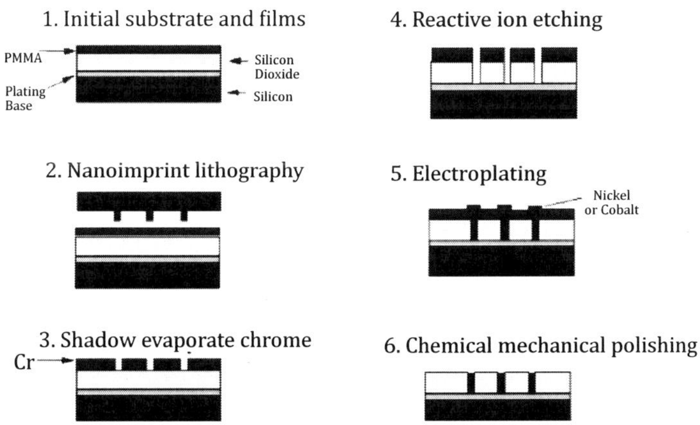

Figure 1.25 Schematic of the quantized magnetic disk fabrication process. Reprinted with permission from Wu et al., 1998. Copyright 1998, American Vacuum Society.

Another fabrication procedure which is much simpler and has the potential to lower the process cost was proposed in 2002

(McClelland et al., 2002). This approach is to pre-pattern a substrate and subsequently deposit a thin magnetic film on it. They used conventional NIL to prepare the nanopatterned SiO_2 substrate and then a multilayer of Co/Pt was covered on it. The patterned densities of 300 Gb/in^2 was achieved by this process.

For the patterned media, the lithography step can be only one and the alignment accuracy is not too strict in contrast to the semiconductor procedure, whereas both sides of the replica must be imprinted to be used as disk drives. Considering that the key for the adoption of patterned media is an inexpensive manufacturing process, nanoimprinting is strongly expected to be one of the most promising methods for it.

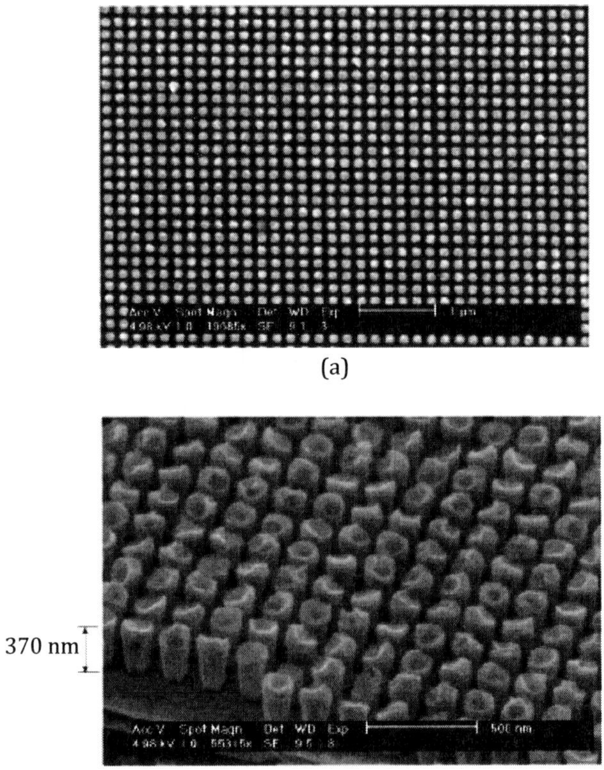

Figure 1.26 SEM images of 18 Gbit/in^2 large area perpendicular quantized magnetic disk. Reprinted with permission from Wu et al., 1998. Copyright 1998, American Vacuum Society.

1.4.2 Optical Devices

For downsizing and weight saving of the general equipments, the miniaturization of optical devices is demanded. Moreover, the progress in the photonic theory and simulation tools, the fabrication size, and accuracy is entering "nanoscale" in recent years. For the mass production of the optical components with nanoscale accuracy, nanoimprint technology is focused and put to practical use. Many optical and photonic device examples such as microlens array, photonic crystal, and waveguide are demonstrated on the basis of nanoimprint technology. In this subsection, two typical devices including our commercialized products will be introduced.

1.4.2.1 MLA (microlens array)

In this part, MLA (microlens array) which is used in an LCD (liquid crystal display) projector will be presented. Projectors have become a standard display tool to be used at the meetings and conferences. Their recent trends are to downsize their dimensions and to improve the quality and brightness of the image. The whole displayed image consists of millions of pixels which are electrically controlled. Especially, in the LCD type projector, the brightness of each pixel is controlled by each TFT (thin film transistor) which is placed for each pixel. Some part of light from the light bulb is shielded by these TFTs and lines to wire them. As a result, the aperture rate of the LCD panel goes down to about 70%.

To improve the efficiency to transfer the light from the bulb to the screen, MLA was proposed. MLA is the array of the microlenses which are attached for each LCD pixel. By attaching MLA and focusing the light, the light can be averted to expose on the TFTs (Fig. 1.27). As a result, the screen can be brighter than that without MLA.

We have commercialized MLA fabricated by using nanoimprint technology. Figure 1.13 described in Section 1.3 explains its brief process flow. The master substrate is made by the photolithography. And then the mold (first generation) is fabricated from the master by electroforming. This electroforming process is repeated and the third-generation mold is used to make the replicas. This repetition of electroforming process can provide a large number of molds from a single master. And this realizes the mass productivity and high reproducibility. Figure 1.28 shows a picture of mold. This mold is used to produce the replica. UV curable polymer is dropped on the quartz substrate and it was pressed on it with UV irradiation. To assemble with the LCD panel, the replicated surface was covered by another polymer. And then its surface is polished to achieve the demanded flatness and thickness. The diameter of each lens is several tens micrometers. Its cross-sectional shape is designed as aspheric lens to realize highest optical efficiency. To replicate its shape correctly, nanometer scale replication accuracy is realized. The number of lenses arrayed in one chip is about a million. And 8 chips are prepared on a mold. As any defects in millions of lenses in a chip are not allowed as a product, the production was conducted in a clean room. These products have been applied to commercialized projectors (Hamada et al., 1995; Aoyama et al., 1996; Aoyama and Yamashita, 1997).

This technique was also used to produce a patterned microlenses for the backlight of mobile phones. Millions of lenses are designed and arranged to convert the light direction on the optimum condition. This improves the efficiency of light use and elongates the battery life time.

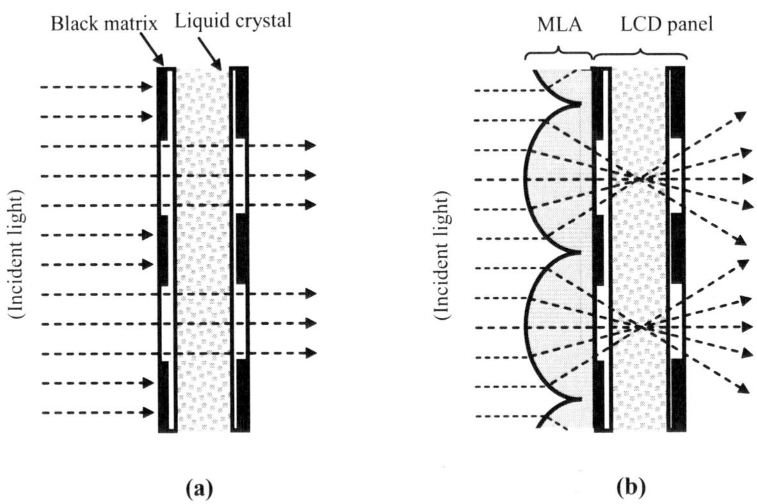

Figure 1.27 Concept images of light transmission through LCD panel without MLA and with MLA.

Figure 1.28 Picture of nickel mold for microlens array.

1.4.2.2 ARS (antireflection structure)

In many optical devices such as optical sensor, display panel, and so on, the anti-reflection coatings consisting of the multiple dielectric (SiO_2, TiO_2, Al_2O_3, etc.) layers are prepared on their panel surfaces to depress the Fresnel reflection. Its principle is based on the optical interferences inside these thin multiple layers and it can counteract the total reflection from the surface. Though this method has been widely used already, it has the drawbacks that the reflection efficiency depends on the wavelength and the angle of incident light. To clear up these drawbacks, the number of anti-reflection layers increases and it results in the increase of process and material cost.

As another way, ARS (anti-reflection structure) which has the nanofeatured patterns on its surface was also proposed and studied (Bernhard, 1967).

Figure 1.29 shows the concept image of the ARS. To avoid the high-order diffraction, the pattern period must be smaller than the half of incident light wavelength. When the nanofeatures are prepared on the surface, the average refraction index gradually changes between the two materials (air and polymer). It follows that the reflection from their boundary can be depressed. The theoretically optimized cross-sectional structure is quadric functional curve (Grann et al., 1995). To fabricate ARS, some traditional processes have been used. The two-beam interference method has been studied to prepare the features. However, the features were basically limited to the sine curve that means we cannot design and optimize the structure. And it is difficult to achieve the reflection coefficient that is smaller than 1.0%. Though the semiconductor procedure such as photolithography was also applied to prepare ARS on silicon substrate, the substrate is not transparent in visible wavelength.

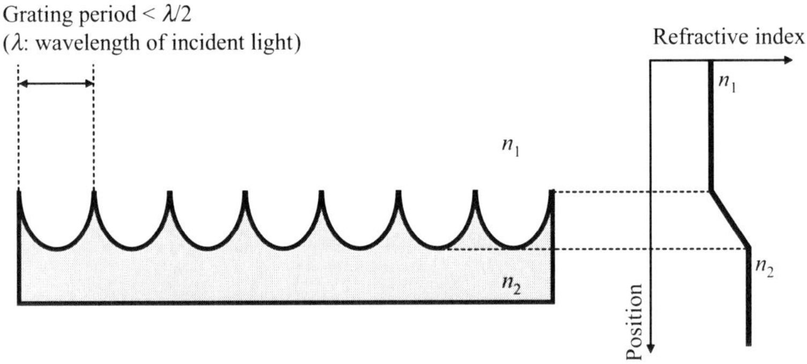

Figure 1.29 Principle of anti-reflection structure device.

To establish a low-cost ARS device with an ideal optical characteristic, we used nanoimprinting method for its fabrication (Nishizaki et al., 2007). The master substrate was made by EB lithography and following D-RIE (deep RIE). And the nickel mold was prepared from this master by electroplating process as described in Section 1.3.2. Figure 1.30 shows the SEM image of the mold surface. The pattern period is 230 nm to avoid the diffraction of visible light. And the pattern height is about 170 nm. Then the patterns are transferred on the replica's surface by using UV nanoimprint method. The UV curable polymer which has the identical refractive index to the substrate (quartz, n = 1.48) was selected to avoid the reflection from the boundary between polymer and substrate. Figure 1.31 shows the SEM image of the replica's surface. The moth-eye structure was formed on the 30 × 40 mm^2 patterned area. The ratio of the caved area to the whole pattern area was about 75% in this result. Figure 1.32 shows the reflection spectrum on the condition that the light enters at a right angle to the surface. The experimental result had a good agreement with that of the simulation. The reflection rate is 0.42% at the wavelength of 450 nm, 0.20% at 550 nm, and 0.31% at 650 nm, respectively. Its specification is almost identical to that of the high-end anti-reflection coating consisting of six dielectric layers. Figure 1.33 shows the comparison of the display panel with and without ARS. Apparently, the reflection was suppressed by attaching ARS on the panel surface.

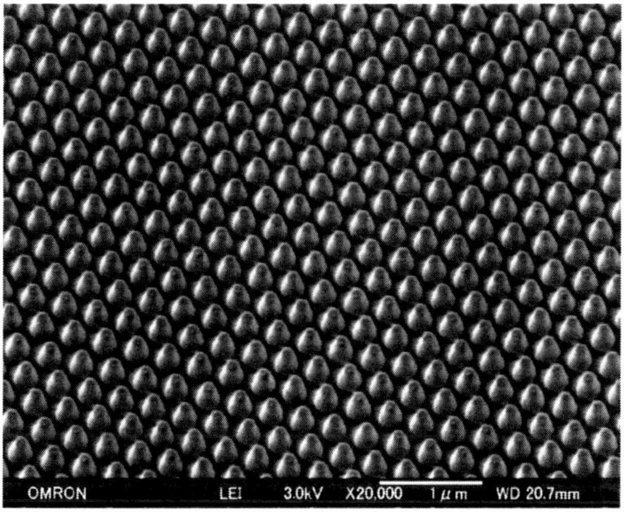

Figure 1.30 SEM image of mold surface on which anti-reflection structure is prepared.

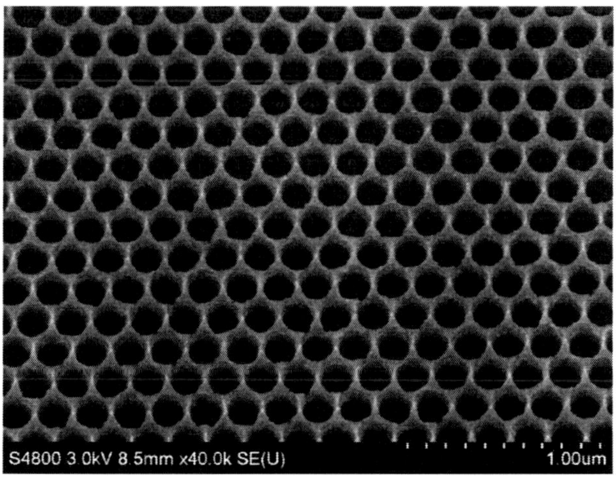

Figure 1.31 SEM image of replica surface on which anti-reflection structure is patterned.

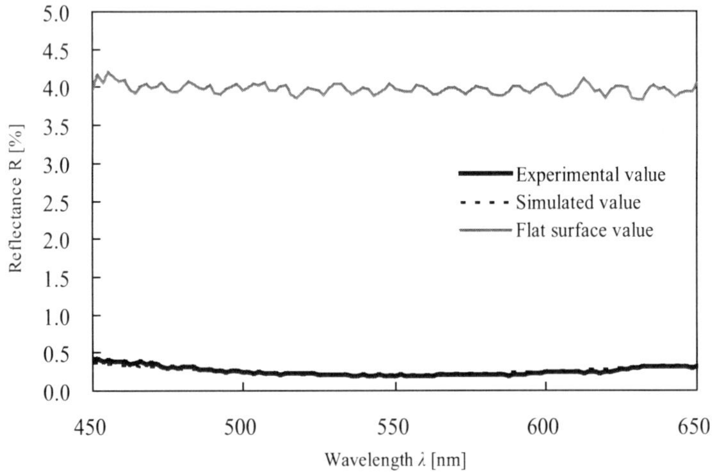

Figure 1.32 Reflection spectra of substrates with/without anti-reflection structure.

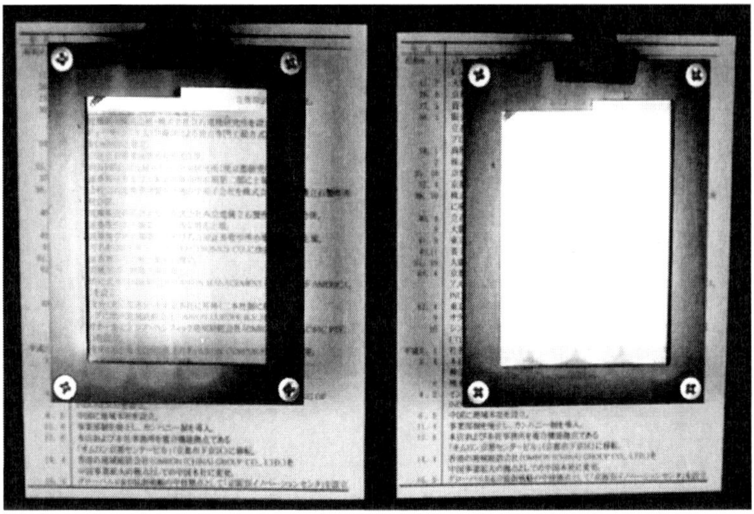

Figure 1.33 Visual effect of anti-reflection structure on display panel.

Not only for the display panel, but the applications for LED are also proposed. LED is expected to be alternative to the conventional lighting tools such as filament lamp and fluorescent light tube because they have long life time (estimation, 40,000 hours). The pointed out bottleneck for pervasion is its light efficiency. By preparing ARS on a p-GaN top cladding layer, the extraction efficiency of the light can be improved as much as five to seven times compared to the identical LED without ARS (Hong et al., 2009). This is also one of the promising applications using nanoimprinting technique.

As another unique and interesting effect of ARS, the surface contact angle can be modified by preparing nanopatterns on its surface. Figure 1.34 explains the principle of contact angle on the nanofeatured surface. The contact angle can be described by Cassie's equation.

$$\cos \theta_f = A_1 \cos \theta_1 + A_2 \cos \theta_2$$

$$(A_1 + A_2 = 1)$$

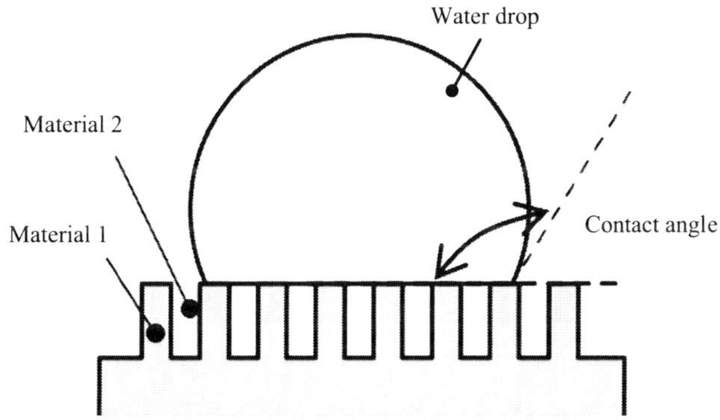

Figure 1.34 Schematic of the contact angle on the nanopatterned surface.

A_1 and A_2 are the surface occupying ratio of two mediums. And θ_1 and θ_2 are the contact angle to the medium_1 and medium_2, respectively. θ_f is the apparent contact angle to these mediums. As the contact angle of air and water is 180 degree, very high contact angle which is difficult to be realized by one solid medium can be achieved by preparing nanopatterns on its surface. The benefit of this surface modification, less dusts can attach and remain on the nanopatterned surface. Figures 1.35 and 1.36 are the demonstration of dust-free effect by a standard dust test regulated by JIS (Japanese Industrial Standards). As the first step, cotton linters are strewn on the panel surface and then they are sucked by the airflow of 0.16 m^3/min. We can see that most dusts on the ARS surface can be removed after the vacuuming, whereas they remain on the flat quartz surface. The removal rate of the cotton linter after vacuuming is 98.5% on the ARS featured surface, although that of the flat surface was 25.4%. This dust-free effect can be a very attractive especially for the applications which will be used on the dusty condition and is difficult to be maintained.

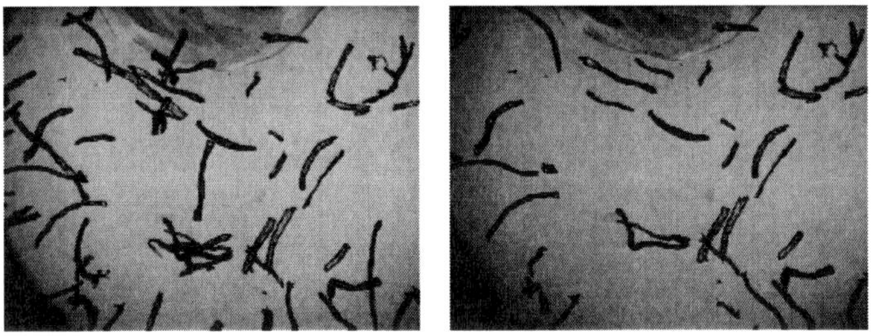

Figure 1.35 Microscopic images before (left) and after (right) sucking the cotton linter on the flat quartz surface.

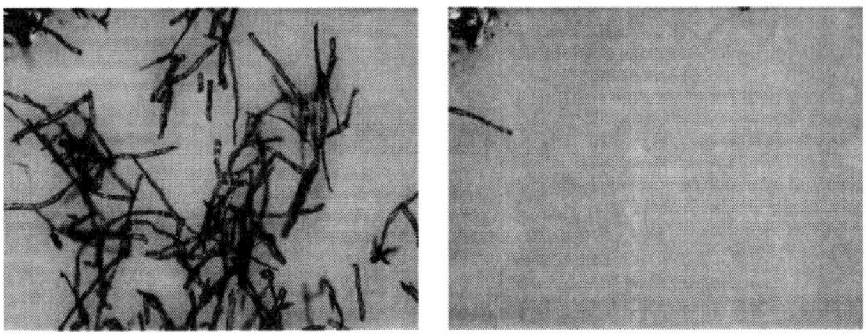

Figure 1.36 Microscopic images before (left) and after (right) sucking the cotton linter on the nanopatterned surface.

1.4.3 Biological Applications

The research trend in biology is to analyze the microconstituent elements such as cell, protein, and DNA, while studying the macrocomponents was mainline in earlier times. As the dimensions of these biological components are from micron to nano order, the micro- and nanopatterned devices play an important role for revealing their functions and controlling their behavior. In this subsection, some recent progresses in this field by the use of nanoimprint technology will be introduced (Truskett and Watts, 2006).

1.4.3.1 DNA and protein arrays by micro- and nanocontact printing

To analyze multiple DNA and protein functions with high throughput, DNA chip and protein chip can be key devices. On these chips, different complementary probes to capture the target molecules are arrayed in advance and multiple interactions are detected simultaneously. To prepare these probe arrays, some methods have been developed. Though direct spotting by the spotting equipment is one of the most common methods, initial investment for the equipment is significant.

On the other hand, unique probe preparation method is proposed by using microcontact printing. As described in Section 1.2.3, this method uses soft molding made from elastomer such as PDMS. On the surface of the micro- and nanopatterns on the mold, the probe materials are coated beforehand (Fig. 1.37). When the mold is pressed on the substrate with appropriate pressuring forces, the top regions on the mold contact on the substrate. As a result, the probe materials on the top of patterns are transferred onto the substrate. The attractive point of this method is that they can be conducted with ease and in low cost by using commercially available precursors. Xiao et al. have prepared a large-scale PDMS stamp consisting of 168 different microstamps on one glass substrate for oligonucleotide synthesis. And they compared the hybridization efficiencies of conventional directly drip-dropped probes and contact printed probes. As a result, no significant difference was observed between these two methods (Xiao et al., 2002).

To achieve a higher resolution, nanocontact printing is studied as an extension of the microcontact printing (Kumar and Whitesides, 1993; Odom, Love et al., 2002; Odom, Thalladi et al., 2002). Some groups realized the structures smaller than 100 nm on printing substrate by using a stiffer elastomeric stamp and by employing high molecular weight inks to avoid diffusions. Hoff et al. demonstrated that 75 nm selective patterning of bioactive proteins was possible with high throughput and high reproducibility by using NIL and fluoropolymer surface passivation. The patterned antibodies and their antigens showed that the biofunctionality and selectivity of probe material remains even after contact printing (Hoff et al., 2004).

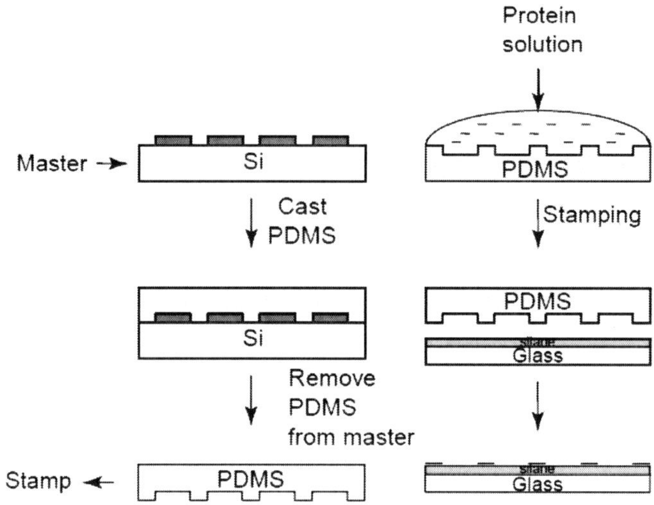

Figure 1.37 Schematic of microcontact printing. Reprinted from Truskett and Watts, 2006, Copyright 2006, with permission from Elsevier.

These contact printing techniques are flexible and can be applicable for many applications such as biosensor, bio-MEMS, tissue engineering, and basic protein research.

1.4.3.2 Tissue engineering

The field of tissue engineering has been focused as one of the main research interests of modern biotechnology. In it, the cells should be cultured at a microenvironment that mimics the extracellular matrix (ECM) to observe its natural behaviors. As ECM consists of nanoscale topographical features, the analysis of the interactions between cell growth and nanopatterned surface can be significant to reveal the mechanism of cell growth and to control it artificially (Curtis and Wilkinson, 1997; Abrams et al., 2000).

Hu et al. (2005) have studied the cell differentiation and growing on the three-dimensional nanostructures made by nanoimprint technique. They prepared some gratings with 10–100 μm half pitch and 200–900 nm height on the commercialized tissue-culture polystyrene (TCPS) and observed cell behaviors on them. Bovine pulmonary artery smooth muscle cells were cultured on the nanoimprinted TCPS plates. F-actin and nucleus in the cells were stained with Oregon Green 488 phalloidin (Molecular Probes, OR) and DAPI (Molecular Probes, CA), respectively, and their morphology and alignment were observed by the fluorescence microscopy. The muscle cell direction on the conventional unpatterned plate prepared as a reference was random and no elongation of cells was observed. On the one hand, the cells on the nanoimprinted TCPS plates whose grating half pitch were 2 μm, 1 μm, and 0.5 μm showed obvious alignments and elongations along the grating direction (Fig. 1.38). Their further study has proved that as smaller the grating half pitch is and as higher the grating height is, the more the alignment and elongation of muscle cells are intensified.

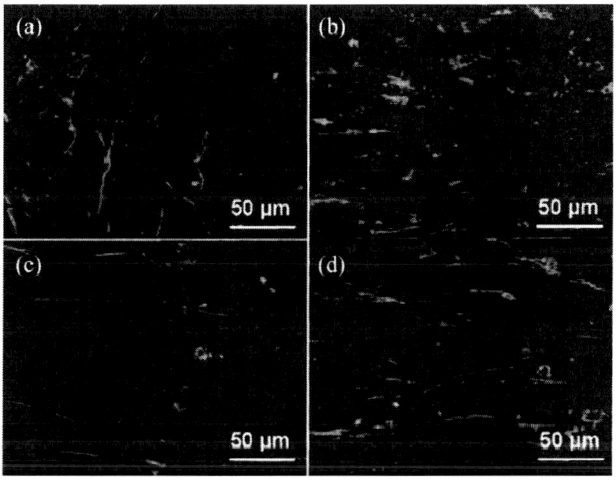

Figure 1.38 Fluorescence micrographs of smooth muscle cells on (a) unpatterned surface, (b) 2 μm half-pitch gratings, (c) 1 μm half-pitch gratings, and (d) 0.5 μm half-pitch gratings. Reprinted with permission from Hu et al. 2005. Copyright 2005, American Vacuum Society.

As another control way of cell growth, Lehnert et al. studied B16 (murine myeloma) cell growth on a plate on which fibronectin substrata was patterned by microcontact printing technique. They have demonstrated that the cell spreading is strongly affected by the density and distance of the substrata. When the space between the substrata dots is smaller than 2 μm, the cells spread as those on a homogeneous substratum. When the space is in the range of 5–20 μm, cells spread and the actin cytoskeleton is elongated between the adjacent dots. At a distance of 25 μm, cell spreading is limited and cells form ellipsoid or round (Lehnert et al., 2004).

These researches have proved that the cell growth and spreading can be controlled by micro- and nanopatterned surface. And they are expected to be used for the effective cell culture of induced pluripotent stem (iPS) cells and so on.

1.4.3.3 Biomolecular imprinting technology

As introduced in Section 1.2.3, direct imprint of biomolecules has been proposed. In our group, we have demonstrated to fabricate the nanopatterns on the polymer surface by the biomolecular nanoimprinting. The detail is as follows:

1.4.3.3.1 *Protein procedure*

Myosin was extracted and purified from back skeletal muscle of rabbit. It was labeled with Cy5-OSu (PA25001; Amersham Bioscience, Uppsala, Sweden). The Cy5-labeled myosin (Cy5-myosin) was diluted to 0.07 mg/ml with an aqueous buffer containing 0.6 M KCl, 20 mM Hepes (pH, 7.0), and 1 mM $MgCl_2$. The myosin solution was dialyzed against an aqueous solution containing 0.15 M KCl, 20 mM Hepes (pH, 7.0), and 1 mM $MgCl_2$ for overnight without stirring. During dialysis, myosin molecules were allowed to aggregate to form self-assembled filaments. The fluorescent myosin filament formation was imaged at an excitation wavelength of 647.1 nm by a total internal reflection fluorescence microscope (Fig. 1.39).

1.4.3.3.2 *Immobilization on a glass substrate*

The aqueous solution containing myosin filaments was dropped on a glass substrate. After several-minutes incubation, the glass surface was cleaned with distilled pure water, and then the surface was dried up under a stream of pure nitrogen gas at room temperature.

1.4.3.3.3 *Nanoimprinting of myosin filaments*

To produce the mold of the substrate on which myosin filaments were immobilized, Ni was deposited on the surface. A 200 nm Ni layer was grown on the substrate by sputter deposition. After that, a 250 μm Ni support layer was further developed by electroforming process. By separating the resulting Ni layers from the glass substrate, the nanoimprinted patterns of myosin filaments were obtained on the Ni mold. The surface of the substrates prepared was imaged with an atomic force microscope (AFM) (NanoScope IV, Veeco Instruments; Woodbury, NY) in gentle tapping mode (Fig. 1.40).

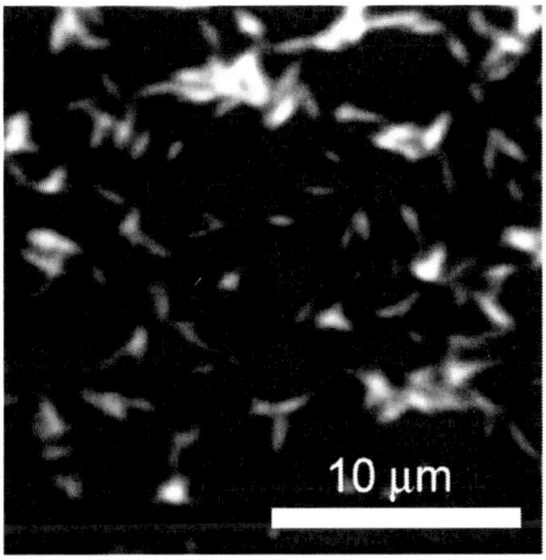

Figure 1.39 Fluorescence image of Cy5-myosin filaments immobilized on a glass substrate.

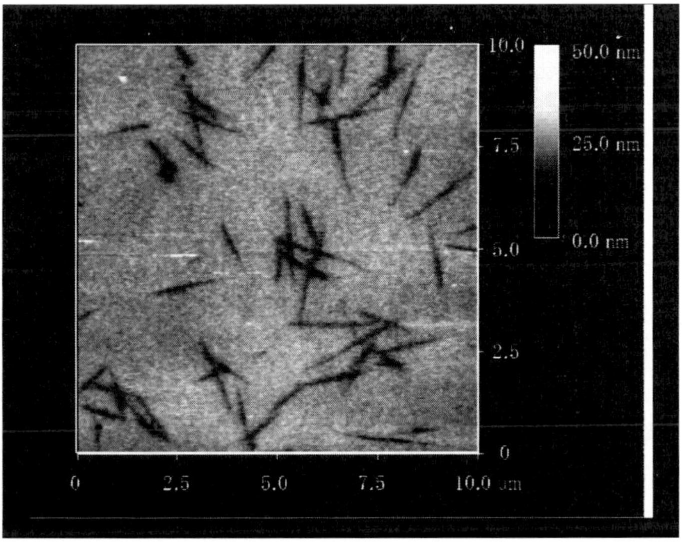

Figure 1.40 AFM image of the mold surface on which patterns of the myosin filaments are transferred.

1.4.3.3.4 *Replication of the nanometer-scale patterns on a PMMA substrate*

The nanoscale patterns transferred from the myosin filaments were developed on a PMMA substrate. The Ni mold was mounted on a glass substrate on which PMMA/ethyl lactate solution was spin coated. The PMMA layer was heated and pressed with the mold. After cooling, the Ni mold was separated from the PMMA substrate and the patterns of myosin filaments were obtained on the PMMA substrate. The surface image obtained by AFM is shown in Fig. 1.41. The lengths and widths were almost constant through the nanoimprint process; therefore, it is indicated that our nanoreplication process was likely to be carried out accurately. Thus, in the present technique, plastic substrates with nanometer-scaled patterns can be massively produced in high precision at a low cost. And by using the biomolecules as the master template, it will be possible to realize the nanoscale and biomimic surface on the solid plastic surface. It has the potential to be applied to the development of the medicine and so on.

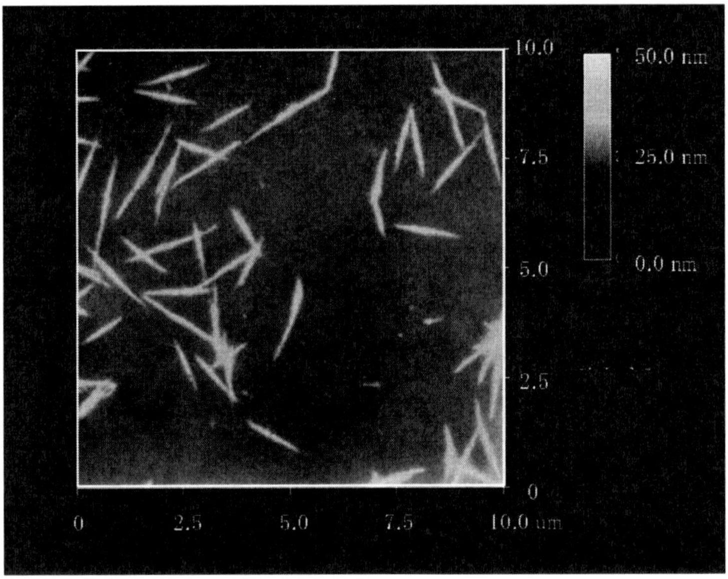

Figure 1.41 AFM image of the replica surface on which patterns of the myosin filaments are transferred.

1.4.4 Energy Devices

According to the increase of the global consumption and the decrease of the residuary fossil fuels, the alternative energy devices such as fuel cells and photovoltaic generation are keenly focused in the recent studies. Some recent studies showed that the nanoimprint technology can also contribute to improve the power generation efficiencies of these devices.

1.4.4.1 Organic solar cells

Among the various energy devices, solar cells are expected as a clean and almost exhaustless power generation tool. Within recent years especially, organic photovoltaic solar cells in which organic electronic materials are used as the light absorption layer have been researched and developed. This method can provide a large area and flexible substrate with easy and low-cost processing as they can be fabricated only by stacking the electrodes and organic materials. However, their power conversion efficiency is still low in comparison to the typical inorganic photovoltaic cells. As a solution to overcome it, some papers demonstrated to prepare the nanostructures on the organic materials. They proved that the efficiencies can be improved by increasing the boundary contact area on these nanostructures (Her et al., 2008; Williams et al., 2008; Baek et al., 2009).

Her et al. prepared well-ordered nanostructures on the titania (TiO_2) which works as the electron acceptor and proved that the efficiency increased 0.16% compared to the conventional flat one. Figure 1.42 shows their relatively unique process flow for preparing the substrate. They used nanoporous alumina made by anodizing Al substrate as a master. Then, PMMA solution was poured on the nanoporous patterns and percolated by heat treatment at 150 degree Celsius. The PMMA mold was separated from the master template by wet-etch process using 1.4 wt% $FeCl_3$/5M HCl solution, followed by 10 wt% NaOH solution. Sol-gel titania layer was prepared by being spin coated over the indium tin oxide (ITO) layer on a glass substrate. Subsequently, the titania surface was nanoimprinted by the PMMA mold with nanopoles. Then the PMMA mold was removed by dipping it in acetonitrile solution and the nanostructured titania could be produced after being baked at 500 degree Celsius for

30 min. They used poly(3-hexylthiophene) (P3HT) as an electron donor and spin coated it onto the nanopatterned titania. The coated thickness of P3HT was about 80 nm. Finally, gold counter electrode (thickness, 100 nm) was deposited on it.

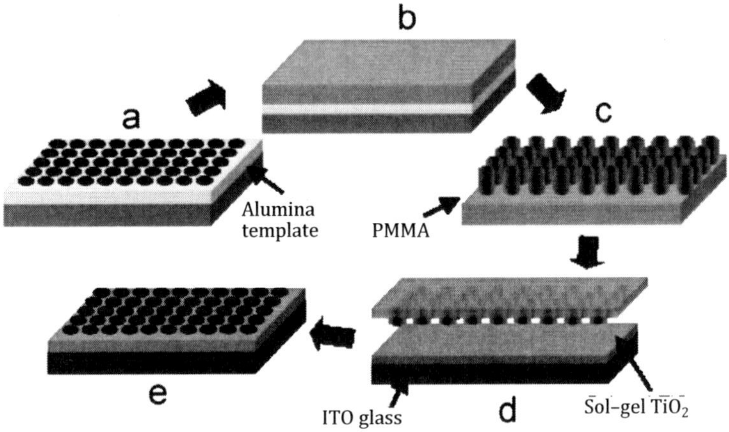

Figure 1.42 Schematic of organic imprint process. Reprinted from Her et al., 2008, Copyright 2008, with permission from Elsevier.

Figure 1.43 shows the SEM images of porous alumina master (a), PMMA mold (b), and nanoimprinted titania (c). The diameter, period, and depth of the nanoporous on titania were 80 nm, 130 nm, and 130 nm, respectively. The power conversion efficiency was estimated from the current density–voltage (*I*–*V*) characteristics with solar simulated light irradiation. As a consequence, the efficiency was 0.16% higher than that of the flat titania surface. This is thought to be contributed by the increment of the titania and P3HT interface area.

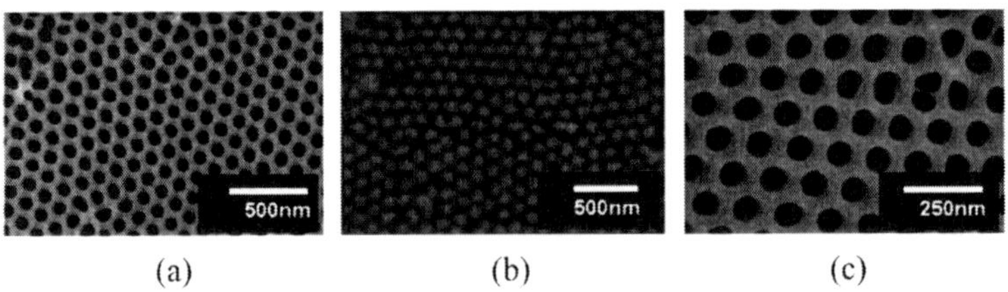

Figure 1.43 SEM images of organic imprint results. Reprinted from Her et al., 2008, Copyright 2008, with permission from Elsevier.

1.4.4.2 Fuel cells

Fuel cell is expected as one of the attractive alternative power generation devices to the conventional batteries. As it converts chemical fuels to electricity directly, the conversion efficiency is very high and it does not exhaust any harmful materials in principle. Several studies have been also reported to produce microfuel cells for the portable usage by using micro-electromechanical systems (MEMS) technology (Kelley et al., 2002; Meyers and Maynard, 2002; O'Hayre et al., 2003; Lu et al., 2004; Kothare, 2006).

In these microfuel cells, the rare and expensive noble metals such as Pt are generally used as a catalyst. To improve the Pt utilization efficiency, Taylor et al. (2007) used nanoimprinting technology on preparing the nanostructured electrodes with Pt thin layer. And they proved that the electrodes with nanoimprinted surface can achieve 18.75 times higher Pt utilization efficiency than those without nanoimprinted structure. Gold layer of 200 nm thickness was deposited on the SiO_2 covering Si wafer with thin Ti adhesion layer. On it, the nanoscale gratings of Pt catalyst with Ti adhesion layer were formed by using NIL. The width, period, and thickness of the Pt bar were 350 nm, 700 nm, and 50 nm, respectively. The nanopatterns are prepared in the region of 13 mm × 4 mm. The characteristics of the fabricated Pt catalyst were electrochemically examined by using cyclic voltammetry. The peak power density of the nanoimprinted electrode was 123 mW/cm^2, which was lower than that of the conventional electrode (410 mW/cm^2). From the point of view of Pt utilization efficiency, however, the former Pt utilization was 15,375 mW/mg, whereas the latter one was 820 mW/mg. This result means that the cost of the electrodes including expensive Pt catalyst can be lowered as much as 18.75 times by using nanoimprint technology on it.

Chapter 2

Nanophotonics and Surface Plasmon Resonance

Light has played an important role in illuminating the life of human beings since ancient times. As its studies progressed, its nature and characteristics were revealed and understood. It has been found that light has three different characteristics from scientific point of view. The first one is the nature of light ray, known as geometrical optics. The second one is the nature of electromagnetic wave, known as wave optics. And the third one is the nature of photon, known as quantum optics. The uniqueness of light is that it includes these different natures in it. They are, however, the scientifically defined different aspects of one thing and related to each other in a very real sense. In this chapter, the first and second ones are included to introduce the basis of nanophotonics and surface plasmon resonance (SPR), which are keenly focused and studied recently. These technologies are expected to illuminate the darkness of the recent unsolved issues such as biological analysis, food and environmental safety, and energy source.

2.1 Nanophotonics

Light is a kind of electromagnetic wave whose wavelength is from about 400 to about 700 nm. It consists of the oscillating electric field

Nanoimprint Biosensors: The Fusion of Nanofabrication, Nanophotonics, and Nanobiology
Takeo Nishikawa and Satoshi Fujita

ISBN 978-981-4316-67-5 (Hardcover), 978-981-4463-42-3 (eBook)
www.panstanford.com

and magnetic field, which are perpendicular to each other. Their relations can be expressed by Maxwell's equations.

$$\nabla \times E = -\frac{\partial B}{\partial t}$$

$$\nabla \cdot E = \frac{\rho}{\varepsilon_0}$$

$$\nabla \times B = \mu_0 \varepsilon_0 \frac{\partial E}{\partial t} + \mu_0 J$$

$$\nabla \cdot B = 0$$

where B is the magnetic field, E is the electric field, J is the electric current density, ρ is the electric charge density, μ_0 is the magnetic permeability, and ε_0 is the dielectric permittivity.

The technologies of light have astonishingly progressed since the laser (light amplification by stimulated emission) was innovated in 1960. The laser has provided the spatially coherent light and many applications are proposed and realized in 1980s. At present, light is used in many fields, as described below, and is necessary for our life and society.

Illuminations

Illumination is the most traditional application of light from the ancient era. The tools for illumination have changed from fire to lamp, fluorescent light, LED (light-emitting diode), and organic electroluminescence.

Fabrications

Light is used to cut various materials. And its exposure is also used to fabricate very small patterns called lithography.

Data Communications, Data Storage

Light can carry digital information with it. And the data can be stored in optical disks and can be read by optical system.

Material Analysis

The constituent substances of a material can be analyzed by optical spectrum known as Raman, FT-IR (Fourier transform infrared), etc.

Energy Sources

Light from the sun is basically the source of our energy. Its use is significantly focused.

The key features of light to be used for these applications are that it can propagate with very high speed and the energy can be focused in a very small area. When used in the macroscopic dimension, light proceeds on a straight course and it can be reflected by mirror and can be refracted by lens. This characteristic as optical ray is the very principle for designing the optical systems.

2.1.1 Light and Diffraction Limit

When the dimension of the optical system is getting smaller and smaller and light is used in the sub-micron region, the optical nature of wave cannot be neglected. The major one is the diffraction of light. The diffraction means the phenomenon in which light propagates with bending when it encounters the obstacles whose size is close to its wavelength. It is caused by the interference by the wave of light and results in the "diffraction limit" that means light cannot be focused on an ultimately small point. The diffraction limit can be expressed as follows:

$$D = \lambda / NA$$

where D means the diameter of focused light, λ is the wavelength of light, and NA is the aperture number of focusing lens (Fig. 2.1).

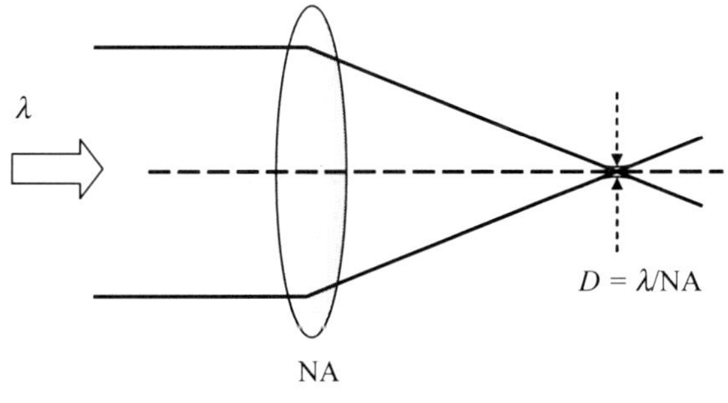

Figure 2.1 Schematic of diffraction limit of light.

This diffraction limit dominates the fabrication limit of photolithography and the data storage density of CD and DVD. To resolve these problems, light sources with shorter wavelength have been developed.

On the other hand, when the object to which light enters is sufficiently smaller than the wavelength, it is known that the electric field formed around the object can be smaller than the diffraction limit. This study is called "nanophotonics" and attracts many researchers as it has the potential to overcome the conventional limit by the diffraction domination.

2.1.2 Nanophotonics and Near-Field Light

Nanophotonics is the study of light in the nanometer scale. When light enters into an object, the atoms in the object form the electric dipole moment as shown in Fig. 2.2. This electric dipole moment generates another electric dipole moment in the neighbor atoms. And the electric field of light can be transferred by the sequence of these electric dipole moments (Fig. 2.3). In the case that the object size is much smaller than the wavelength of the incident light, the distribution of the electric dipole moments is not affected by the incident wavelength since the phase shift of the incident light can be neglected in the very small region. In consequence, the electric field formed around the object is not decided by wavelength of incident light, but determined by the shape and size and material of the object. This means that the near field can be smaller than the diffraction limit of light.

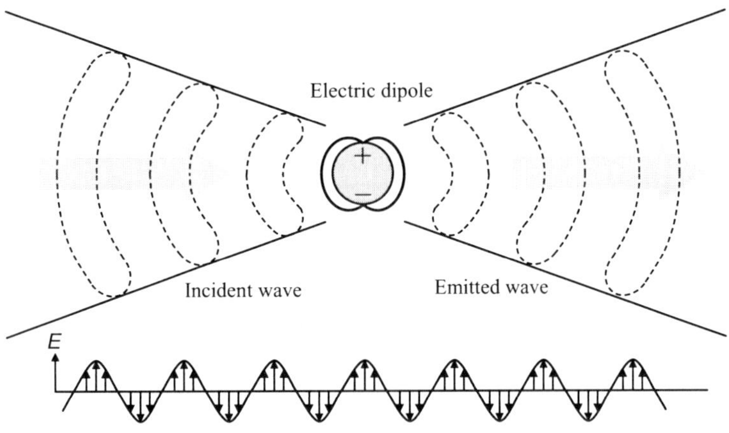

Figure 2.2 Dipole emission schematic image.

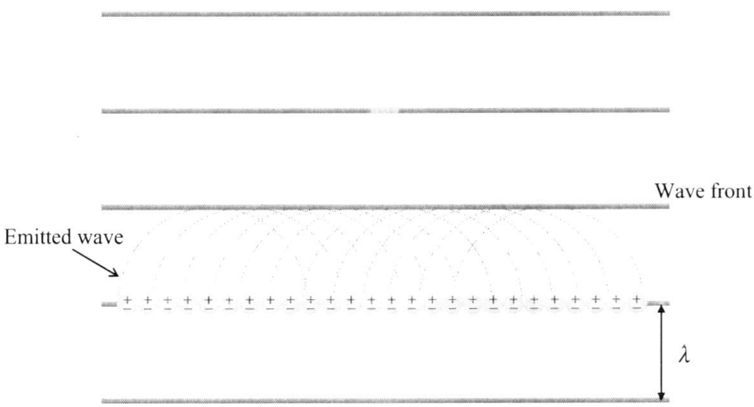

Figure 2.3 Wave propagation in a dielectric material.

By using this near field, light can be concentrated in a very small area (less than several tens of nanometers). It is proposed that this can be used to realize the optical integrated circuit composed of the nanophotonic devices. Whereas the conventional electric integrated circuit uses electrons as its operational media, the optical integrated circuit uses light as its operational media. And it has the potential to provide an ultra high speed and low energy consumption and low heat evolution device. It consists of the light source and amplifier, optical switch, and detector. When using the conventional optical devices for them, the dimension of the circuit must be much larger than that of the current electric circuit as their sizes are dominated by the diffraction limit of light. Therefore, the nanophotonic devices are expected to solve these problems and to realize the supreme integrated circuit.

In case that the object is made of the metal, the near field can be extensively enhanced by the generation of localized surface plasmon resonance (LSPR) that will be described in another section. By these new and unique technologies, many applications of the nanophotonics are proposed and developed in this decade.

2.1.3 Scanning Near-Field Optical Microscope

As the near field stays in only the vicinity of the object, detecting it from far region is difficult. To detect it, another object which works as a probe needs to be approached to it. When the probe gets close to it, the electric dipole moments of the near field couple to the probe and some parts of it propagate along the probe. As a result, the near field can be detected by a macroscopic region. This system is known as SNOM (scanning near-field optical microscope) or NSOM (near-field scanning optical microscope) and commercialized by some companies. SNOM can be divided roughly into two types depending on the forms of probe used. The first one is a "scattering type," which uses a sharply pointed out metal probe. And the second one is an "aperture type," which has an aperture whose size is smaller than the wavelength of light on its top. Table 2.1 shows their diagrams and features.

Table 2.1 Comparison table of SNOM types

	Scattering type	**Aperture type**
Image	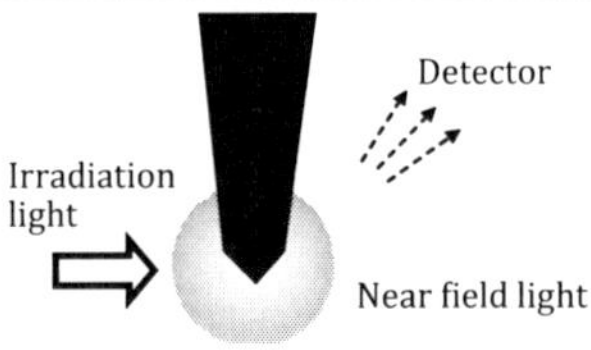	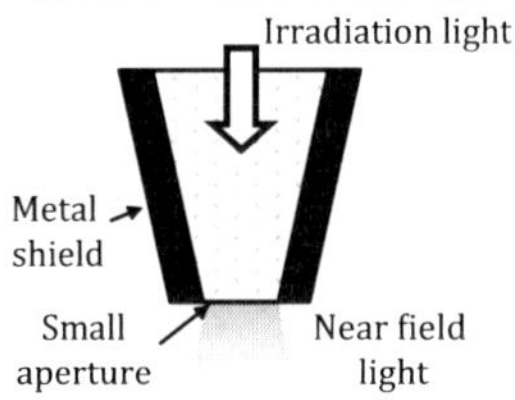
Advantages	• High resolution (~10 nm) Possible to follow the small irregularity on the surface	• Low noise
Drawbacks	• Scattering light from other places than the probe top can be detected as noise.	• Low resolution (>100 nm)

Though the scattering type has a high spatial resolution, it detects not only light from the top of probe, but also light scattered from the other components. Therefore, when using this type, the irradiation angle of incident light, the orientation of the detector, and so on must be taken care. On the other hand, the aperture type can detect only the near-field light as it detects and propagates it directly through the aperture and fiber. And it can also emit the irradiation light from the aperture of probe that realizes the excitation-emission characteristics analysis in nanoscale. The spatial resolution of the aperture type is, however, lower than the former because the aperture needs some area to collect the enough light to detect. Here, the aperture type SNOM instrument "NFGP120" (Fig. 2.4) developed by Nihon Bunko Corporation will be introduced as an actual example.

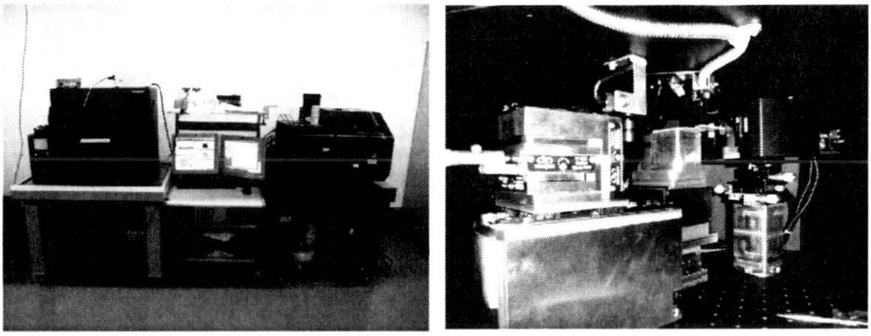

(a) (b)

Figure 2.4 The pictures of SNOM: (a) exterior view and (b) inside measurement chamber.

The fundamental system diagram is illustrated in Fig. 2.5. Light from the source (wavelength, 532 nm; power, 30 mW) is guided to the aperture at the top of the probe by the optical fiber. The excitation light is emitted from the aperture and irradiated on the target sample. The emission light from the sample is collected by the aperture and guided to the spectrometer through the fiber. The excitation and emission lights are divided by the beam splitter on the way to the detector. The spectrometer consists of the grating and a cooled-CCD (charge coupled device) to detect the faint light. The distance between probe and sample is regulated by measuring the shear force between the end of near-field probe and the sample. The probe is scanned along x- and y-directions to acquire the mapping data. In NFGP120, there is a wide space under the sample stage for setting the optical system to irradiate light from the backside of the sample substrate.

What dominates the performance of SNOM is the performance of its probe. Figure 2.6 indicates an example of SEM image of the aperture type probe. There is a fiber aperture at the center of probe top which is surrounded by metal mask. The smaller the aperture diameter is, the higher the space resolution can be. The acquirable light, however, decreases rapidly when the aperture size gets smaller. Therefore, the probe with 100–300 nm aperture diameter is generally used to observe the near-field light.

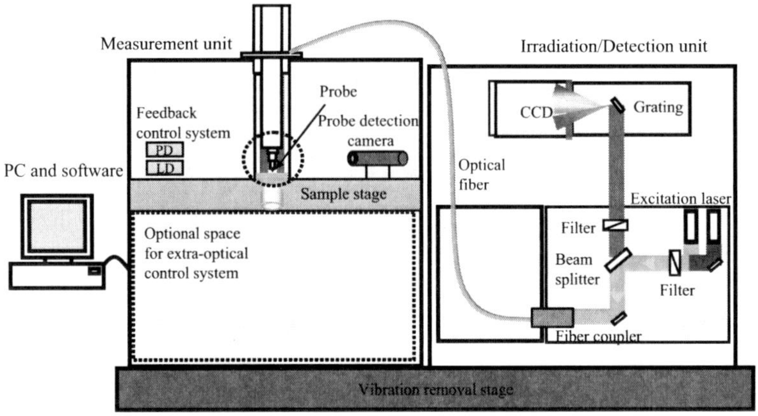

Figure 2.5 Block schematic of SNOM system.

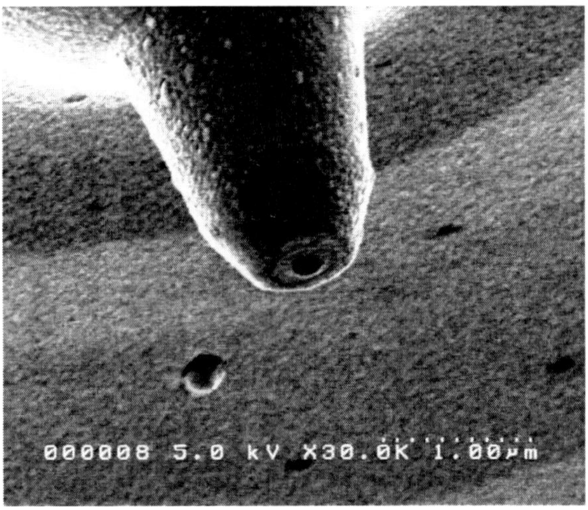

Figure 2.6 SEM image of aperture type probe for SNOM. The aperture diameter is about 100 nm.

Figure 2.7 shows an example of topography and optical intensity measured by NFGP120. The sample was prepared by using electron beam lithography and lift-off process. The gold dot array with the period of 180 nm, diameter of about 100 nm, and height of about

100 nm was prepared on a silicon wafer. Figure 2.7(a) shows the topography result and Fig. 2.7(b) shows the intensity mapping of integrated spectrum between 500 nm and 580 nm. The irradiation wavelength is 532 nm, the exposure time is 1 sec, and the aperture diameter of the probe is 120 nm. We can apparently see that the intense area of the spectrum corresponds to the high position of topography. The spectrum data on the intense and weak points is shown in Fig. 2.8. The collected light from the gold dot position is a little stronger than that from the gap position especially at around 540 nm. This is assumed to be caused by the gold dots that generate the localized SPR and the incident electric field was intensified by it. By using SNOM equipment, we can observe the optical and photonic characteristics in nanoscale that cannot be analyzed by the conventional systems.

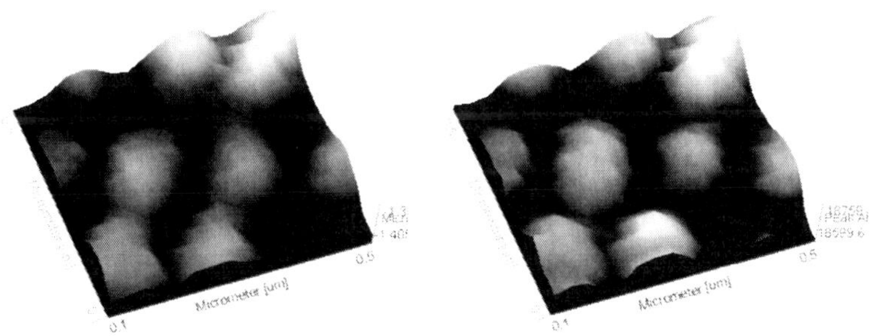

(a) (b)

Figure 2.7 SNOM measurement results of (a) topography and (b) optical intensity.

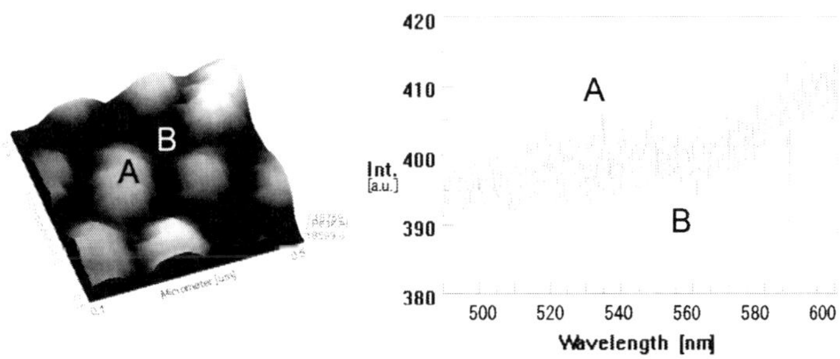

(a) (b)

Figure 2.8 SNOM results of (a) space map and (b) spectrum data.

2.2 Surface Plasmon Resonance

As described in the previous section, light is the propagating oscillation of the electromagnetic field and its distribution is changed by the encountering object. When the object is made of the metal, the phenomenon becomes more complex than that in the dielectric materials as numerous free electrons exist in it. The metal free electrons are easily affected by the incident electric field and can provide some unique characteristics. One of them is propagating SPR, which is the resonance between the incident light and the free electrons on the flat metal surface. When the size of metal is much smaller than the wavelength, LSPR can occur on a specific condition. These phenomena are attracting intensive attention in optics and nanophotonics field and many applications using them are proposed and demonstrated in these years.

2.2.1 Plasma Oscillation in and around Metal

"Plasma" is generally used for the condition in which the ionized gas keeps neutrality in total. In metal, the free electrons and positive metal ions keep neutrality, too. By this analogy, the neutral condition in the metal is also called "plasma." Plasma oscillation is a movement of the free electrons in the "bulk" of the conducting media such as metals. Considering the neutral plasma condition in metal, there are positively charged metal ion and negatively charged electrons in it. When free electrons are perturbed and activated collectively, the Coulomb force pulls them back. This force results in an oscillation of free electrons on an eigenmodal frequency decided by the density of free electrons. This plasma frequency ω_p can be depicted by the below equation:

$$\omega_p = \sqrt{\frac{4\pi n_e e^2}{m}}$$

where n_e is the density of electrons, e is the electric charge, and m is the mass of the electron.

This plasma frequency is an important parameter to decide the optical characteristic of the metal. When light with the frequency lower than ω_p enters the metal, the opposite electric field is produced by the movement of the metal free electrons following the oscillation of the light electric field, as a result light cannot go into it and is reflected. When the frequency of light is higher than ω_p, the free electrons cannot follow the oscillation frequency of the incident light and, as a result, light can penetrate it.

Near the surface of the metal, the free electrons are also affected by the neighboring dielectric material. And their movement differs from that in the "bulk" condition. This resonance mode is called SPR. On the boundary between the flat metal and dielectric material, the so-called propagating SPR mode exists. The detail of propagating SPR will be introduced in the next part.

2.2.2 Propagating Surface Plasmon Resonance

Propagating SPR is an electromagnetic wave which propagates along the metal surface. It is caused by a resonant interaction between the incident light and the free electrons in metal which occurs at the metal/dielectric (metal/vacuum) interface. The dispersion relation equation can be expressed as follows:

$$k_{sp} = \frac{2\pi}{\lambda} = \frac{\omega}{c}\sqrt{\frac{\varepsilon_m \cdot \varepsilon_s}{\varepsilon_m + \varepsilon_s}}$$

where k, λ, ω, and c express the wave number, wavelength, angular frequency, and light speed, respectively. ε_m and ε_s are the permittivity of the metal and the neighboring material which are the functions of ω. ε_m can be calculated by using Drude model or Lorentz–Drude model, generally. As an example, a schematic image of the dispersion relation of gold/water interface is represented in Fig. 2.9. To generate this SPR in actual experiment, external excitation by the incident light is necessary. The dispersion relation of light in a dielectric material can be expressed as follows:

$$k_l = \frac{2\pi}{\lambda} = n \cdot \frac{\omega}{c}$$

where n expresses the refractive index of the material in which light proceeds. According to Fig. 2.9, it will be noticed that there is no intersection point between the SPR mode and the light mode. This means that the SPR cannot be generated by the general propagating light.

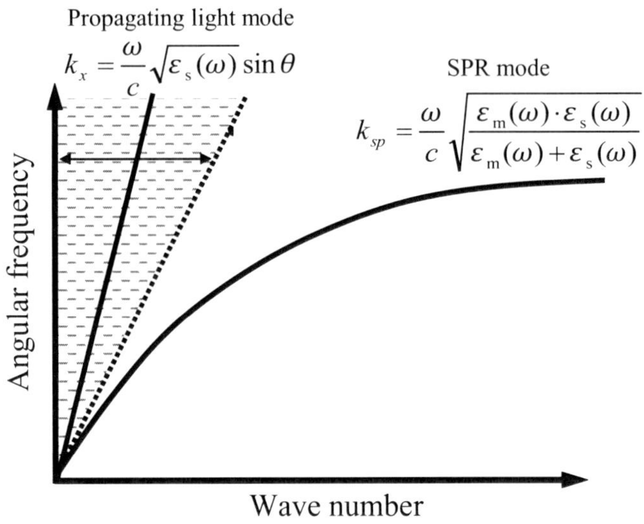

Figure 2.9 Schematic image of dispersion relation of propagating light and SPR mode.

It is said about the origin of SPR excitation discovery that Wood found the dark line in the spectra from the optical grating depending on the concentration of glycerin solution in 1902 (Wood, 1902). This dark line was called anomaly and was out of favor with spectrum researchers. This was revealed to be caused by a specific wavelength light that was absorbed by SPR occurrence on the metal grating surface (Fig. 2.10). This is the first excitation method of propagating SPR. The wave number of the diffracted light shifts $m{\cdot}k_g$ ($k_g = 2\pi/\Lambda$) by the grating with the period of Λ ($m = \pm 1, 2, 3, \ldots$). Figure 2.11 shows its dispersion relation. It shows that there exists an intersection between the SPR and grating incident light. The resonant condition can be expressed by the next equation.

$$k_{sp} = k_l + m \cdot k_g$$

Grating excitation method does not require a complex optical configuration when the grating patterns can be prepared on the metal surface.

As another SPR excitation method, it was found that light wave in the total internal reflection (TIR) can couple to the SPR mode. The dispersion relation of TIR light can be expressed as the next equation.

$$k_{\text{TIR}} = \frac{2\pi}{\lambda} = n \cdot \frac{\omega}{c} \cdot \sin\theta$$

where θ means the incident angle of light to the TIR interface. This can be understood that the wavelength at the TIR can be $\lambda/\sin\theta$ as shown in Fig. 2.12. This relation can be expressed as dotted line in Fig. 2.13. We can see that the intersection point exists between SPR and TIR light mode. On the condition that satisfies the next equation, SPR can be generated by the incident light.

$$k_{\text{sp}} = k_{\text{TIR}}$$

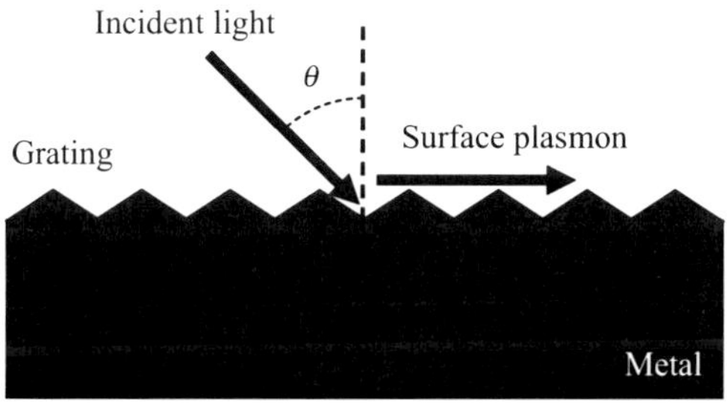

Figure 2.10 Generation of surface plasmon resonance on a grating.

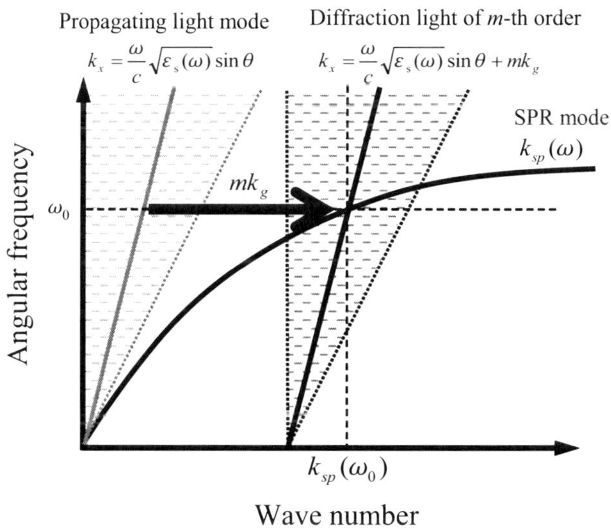

Figure 2.11 Dispersion relation of surface plasmon resonance coupled with diffraction light.

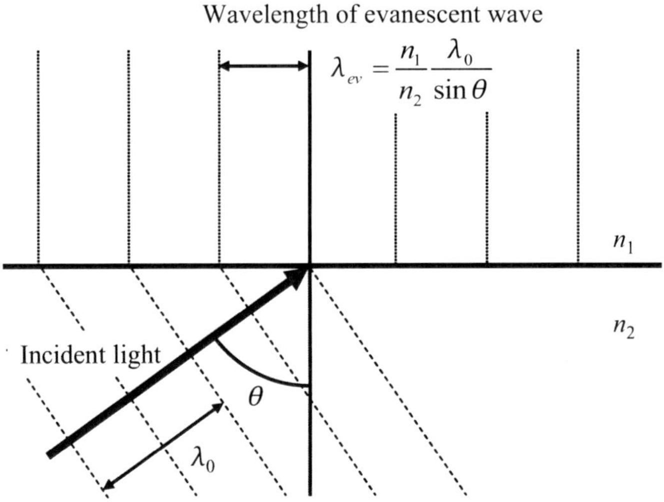

Figure 2.12 Schematic of attenuated total internal reflection.

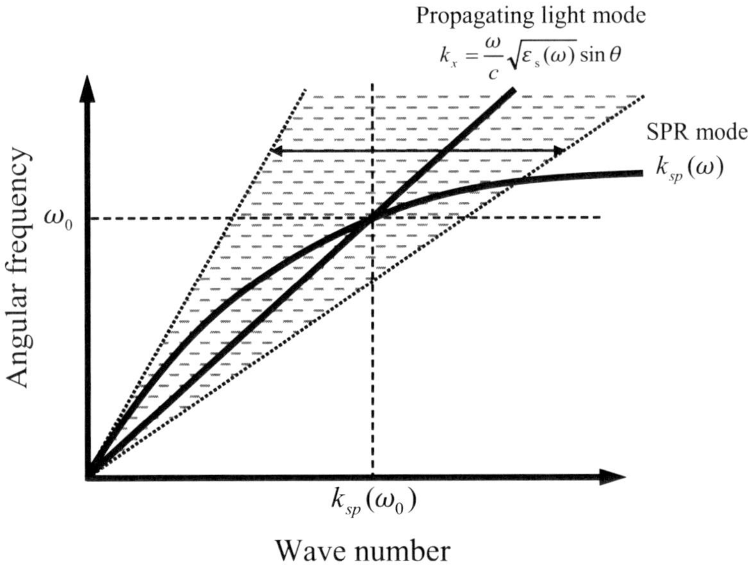

Figure 2.13 Dispersion relation of surface plasmon resonance coupled with attenuated total internal reflection light.

In the actual experiment, Otto (1968) has first proved that SPR can be generated by using the optical prism. In Otto configuration, the bulk metal is set up to face its surface to the prism reflection surface with a gap of a few hundred nanometers (Fig. 2.14). Then light enters from the bottom side of the prism for generating an ATR (attenuated total internal reflection). The evanescent field on the resonant condition couples to the SPR mode on the metal surface and generate the propagating SPR. The reflection light on that condition decreases as its energy is transferred to the generation of SPR that propagates along the metal surface. Kretschmann demonstrated another configuration to generate SPR by using an evanescent field (Kretschmann and Raether, 1968). In Kretschmann configuration, the thin metal layer is directly attached onto the prism surface. And the evanescent field on the prism surface couples to the SPR mode on the metal surface (Fig. 2.15). Since his method does not require precise control of the gap between the prism and metal surface, it is commonly used for SPR experiment recently. Figure 2.16 shows schematic of an actual experimental set up for observing SPR. In this system, the collimated monochromatic (wavelength, 633 nm) light is focused by using cylindrical lens (focal length, 300 mm) and entered into the optical prism (refractive index, 1.514). The focus of cylindrical lens is set at the ATR surface. Then light is reflected at the back surface of a thin gold layer (thickness, 50 nm) on the prism surface and proceeds to another collimator lens. The glass substrate with the gold film is attached on a prism with the matching oil. Light is detected by CCD camera whose detector cells correspond to the angular information of incident light. There are other methods to generate and observe SPR such as using a galvanometer mirror to scan the incident angle of light. Furthermore, the SPR curve can be observed by using white light to generate SPR and detecting the reflection light by a spectrometer as the above equation can be satisfied at a specific wavelength at a fixed incident angle. About the kind of metal, the SPR can be generated by silver or aluminum. Gold is, however, most commonly used in the experiment especially for the sensor applications as it is much more chemically stable than other metals. Figure 2.17 is the experimental result which shows the reflection as a function of incident angle. The reflection local minimum, also called "dip," was observed at about 73.5 degrees due to the generation of SPR.

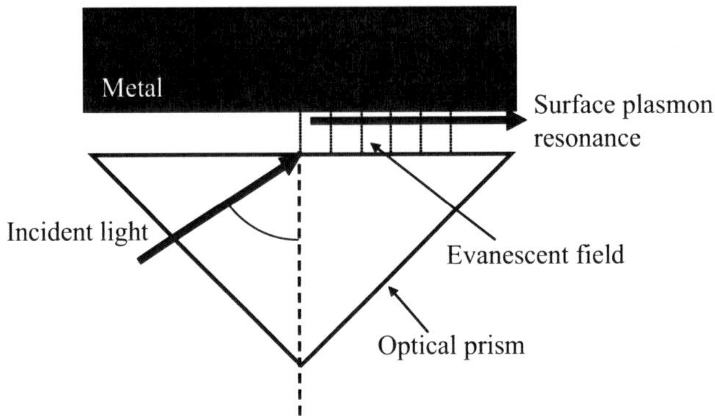

Figure 2.14 Generation of surface plasmon resonance on Otto configuration.

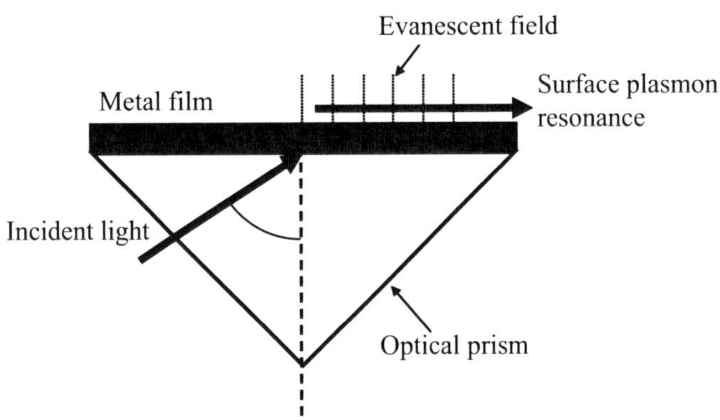

Figure 2.15 Generation of surface plasmon resonance on Kretschmann configuration.

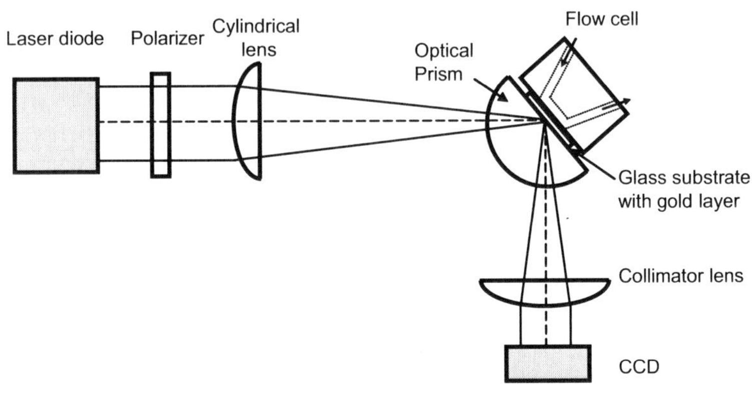

Figure 2.16 Optical experimental set up for surface plasmon resonance.

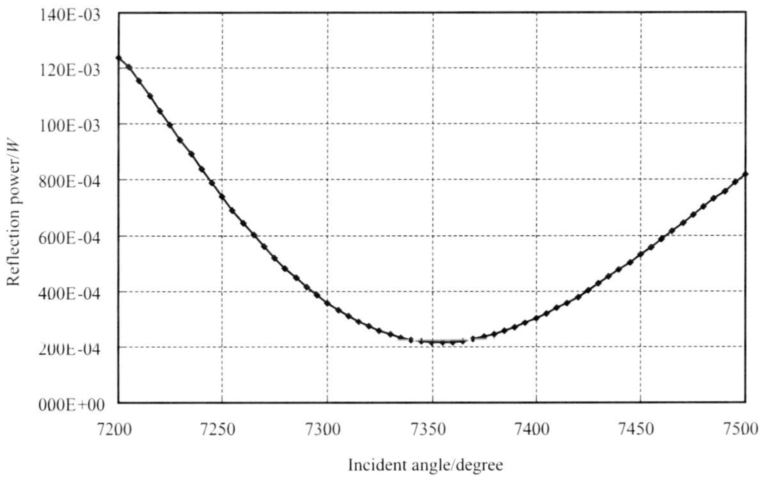

Figure 2.17 Experimental reflection curve as a function of incident light angle.

2.2.3 SPR Theoretical Calculation

For the development of SPR applications, its theoretical calculation is meaningful and helpful in analyzing the results and deciding the experimental parameters. The basic principle for simulating the general propagating SPR is to calculate the reflection coefficient from the multilayered substrate including the complex refractive index of materials. The decrease of reflection coefficient by SPR occurrence is accounted by the imaginary part of metal refractive index as the absorbance in it. To solve them, transfer matrix method (TMM) (Born and Wolf, 1975; Chilwell and Hodgkinson, 1984), which is one of the most common formulae, is briefly introduced in this section.

TMM uses matrix form that is extensively useful for the steady solution to Maxwell's equations subject to the boundary conditions on the multilayer stack. Consider the stack of N layers as illustrated in Fig. 2.18. The layers are numbered from 1 to N, whereas the layer interfaces are numbered 1 to $N - 1$. The temporal and spatial dependence of the waves propagating in the positive y-direction can be expressed as $\exp\{ik(\pm\alpha x + \beta y) - i\omega t\}$. Here ω is the radian angular frequency and k is the vacuum wave number of incident light. By defining the incident angle as θ, α, and β can be represented as follows:

$$\alpha = n \cos \theta$$

$$\beta = n \sin \theta$$

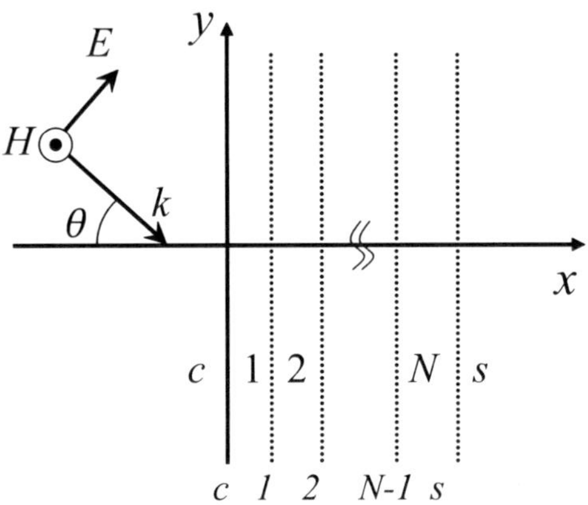

Figure 2.18 Schematic model of basic transfer matrix method.

In this simple coordinated system, there is material variation along only x-direction. Furthermore, the propagating SPR can be generated only by TM polarized incident light since TE component of light does not affect the free electrons along SPR wave. Maxwell's equations can be extremely simplified as a consequence. On this limited condition, three following field components are zero.

$$E_z = H_x = H_y = 0$$

The amplitude of the nonzero components H_z, E_y, and E_x are assigned to U, V, and W, respectively. According to Maxwell's equations, the amplitudes of these field components are related by

$$V = \frac{\gamma}{ik\alpha}\frac{\mathrm{d}U}{\mathrm{d}x}$$

where γ is $z_0 \cos\theta/n$, $z_0 = (\mu_0/\varepsilon_0)$ is the impedance of free space and n is the refractive index of nonmagnetic medium in which the wave is traveling. W is the amplitude of the E_x which is normal to the interfaces and is proportional to U, whereas U and V are tangential to the interfaces. Therefore, a total field can be adequately specified by the vectors U and V.

$$W = \frac{\beta\gamma}{\alpha} U$$

$$U = \frac{1}{ik\gamma\alpha} \frac{\mathrm{d}V}{\mathrm{d}x}$$

A field transfer matrix M_j relates the field amplitudes U_j and V_j at x_j to the corresponding amplitudes at a neighboring point x_{j-1}.

$$\begin{pmatrix} U_{j-1} \\ V_{j-1} \end{pmatrix} = M_j \begin{pmatrix} U_j \\ V_j \end{pmatrix}$$

The form of the matrix M_j can be derived by solving Maxwell's equations in terms of U and V.

$$M_j = \begin{pmatrix} \cos\Phi_j & -\frac{i}{\gamma_j}\sin\Phi_j \\ -i\gamma_j \sin\Phi_j & \cos\Phi_j \end{pmatrix}$$

where $\Phi_j = k\cdot\alpha_j\cdot(x_j - x_{j-1})$ is the phase thickness of the layer.

The transfer matrix M whose consisting elements are m_{11}, m_{12}, m_{21}, and m_{22} for a stacked N layers is given by the product of the respective transfer matrices for the individual layers.

$$M = \prod_{j=1}^{N} M_j = \begin{pmatrix} m_{11} & m_{12} \\ m_{21} & m_{22} \end{pmatrix}$$

Suppose a plane wave with TM polarization in the cover is incident on the stacked layer, then the reflection coefficients r can be derived by relating the amplitudes of the incident, reflected, and transmitted traveling fields.

$$r = \frac{\gamma_c m_{11} + \gamma_c\gamma_s m_{12} - m_{21} - \gamma_s m_{22}}{\gamma_c m_{11} + \gamma_c\gamma_s m_{12} + m_{21} + \gamma_s m_{22}}$$

Finally, the reflectance R can be expressed by the following equation:

$$R = |r|^2$$

This equation can be calculated by relatively simple computer program. We have constructed this calculation software based on Excel Macro program. Figure 2.19 shows an example calculated by this program on the conditions that a gold layer of 50 nm thickness is deposited on a glass (n = 1.49) substrate and its bare surface is facing at water (n = 1.33) and that the incident light (wavelength, 633 nm) enters from the glass substrate side with various incident angles. The dip of the reflection curve was observed at about 73.5 degree and this result has a good agreement with the experimental result in Fig. 2.17.

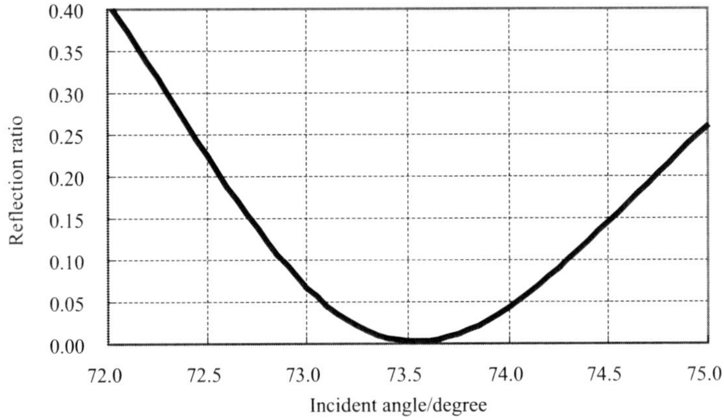

Figure 2.19 Calculation result of the reflection ratio as the function of the incident angle of light. The calculation algorithm is based on transfer matrix method.

2.3 Localized Surface Plasmon Resonance

In the 17th century, Andreas Cassius (1605–1673) invented a pigment called "purple of Cassius" and Johann Kunckel (1630–1703) accomplished a ruby-colored stained glass with it. In 1857, Michael Faraday made a red-colored colloidal solution from deoxidized gold ions by adding carbon disulfide solution to $HAuCl_4$ (Faraday, 1857). After that, Richard Zsigmondy proved that the purple of Cassius resulted from the color of gold colloid and he received Nobel Prize in 1925 because of these achievements.

The color of gold colloid derives from the absorption of light with a specific wavelength range. This is caused by the resonance between the incident light and the free electrons in the gold particle. Compared to the propagating SPR on a flat metal surface, this phenomenon is called "localized surface plasmon resonance (LSPR)." LSPR occurs on the metal nanosized structures such as colloids and does not propagate and confined in the nanoscale region. As a result of this confinement, the resonant electric field is localized in a smaller area than the diffraction limit of light. Due to these unique characteristics and the advances in the nanotechnology, the studies and applications of LSPR are reported in recent years.

2.3.1 Theory of Localized Surface Plasmon Resonance

When the metal colloid is a complete sphere located in a uniform dielectric material, the LSPR mode can be solved from Maxwell's equations. Furthermore, in case that the diameter of the colloid is much smaller than the wavelength of the incident light, the phase shift inside the colloid can be negligible and the Maxwell's equations can be simplified through means of an electrostatic approximation. By solving the Laplace equations in terms of the scalar potential and applying the boundary conditions (Hayashi, 2002), the polarization coefficients of metal colloid can be expressed as follows:

$$\alpha = 4\pi r^3 \frac{\varepsilon_1(\omega) - \varepsilon_2}{\varepsilon_1(\omega) + 2\varepsilon_2}$$

where r is the radius of the metal colloid and $\varepsilon_1(\omega)$ and ε_2 are the permittivities of the metal and ambient material, respectively. When the real component of the denominator of the above equation close to zero, i.e., $\mathrm{Re}[\varepsilon_1(\omega) + 2\varepsilon_2] = 0$, the polarization becomes enormous. This means that a resonance occurs on this condition. The frequency satisfying this condition is called Fröhlich frequency (Fröhlich, 1949). The resonant conditions of the gold and silver derived from the above equation are about 510 nm and 360 nm, respectively.

The Fröhlich mode can be understood by considering the electron oscillation under the effect of an external electric field. Figure 2.20 shows the image of metal colloid in which the polarization occurs by the incident light. When the external electric field enters into the particle, the free electrons inside it are affected and move toward positive direction of the external electric field. As a result, the collection of these free electrons forms a new inner electric field against the external field. The free electrons feel the force to negate the inner polarization. When the oscillation of the electric field matches this movement and the free electrons are accelerated to return to their first position not only by the inner polarization but also by the external field, the amplitude of the oscillation is intensified. This repetition of intensified oscillation of the free electrons results in resonance called LSPR.

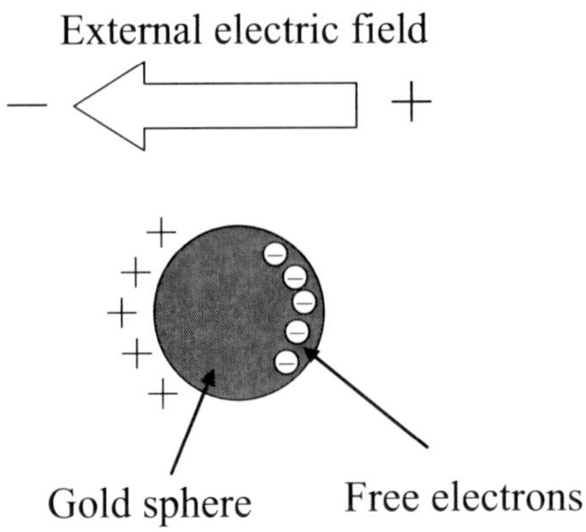

Figure 2.20 Schematic of the free electron polarization inside the metal nanosphere by the external electric field.

2.3.2 Various Modes of LSPR

Basically, the study of LSPR has been started about single spheric nanocolloid. In recent researches, however, many LSPR modes on particles with various shapes are studied and analyzed. In this section, some typical example will be introduced (Yamada, 2009).

2.3.2.1 Gold nanorods

Other than the sphere colloids, gold nanorod is one of the most popular and focused shapes for LSPR devices (Vial et al., 2007). It exhibits orient-dependent optical responses as shown in Fig. 2.21. The absorbance peak at about 520 nm corresponds to the resonance on a short axis of nanorod (transverse mode) and one at about 800 nm corresponds to the resonance on a long axis of nanorod (longitudinal mode). Stephan and Mostafa (1999) have calculated the absorbance spectra by applying Mie theory on nanorod shape with some approximations. Figure 2.22 shows its result. It is clarified that the resonance peak of longitudinal mode mainly depends on its aspect ratio (long axis/short axis).

Some fabrication procedures for nanorods are proposed and established. The first one is electrochemical method that requires

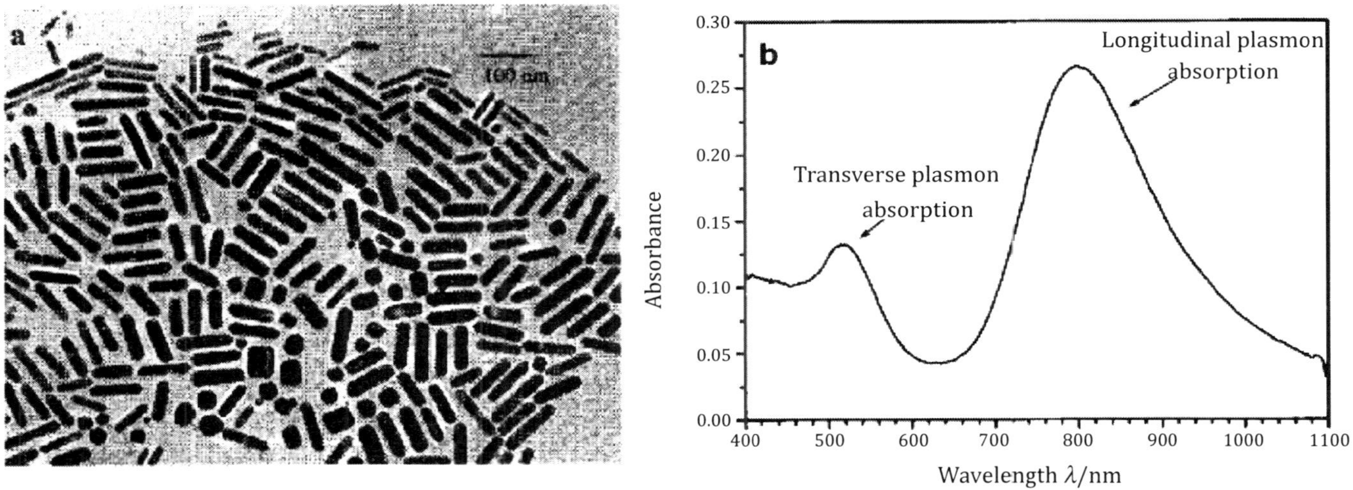

Figure 2.21 SEM image (left) and absorbance spectrum (right) of gold nanorod. Reprinted with permission from Stephan and Mostafa, 1999. Copyright 1999, American Chemical Society.

an electrochemical solution composed by interfacial active agent, acetone, cyclohexane, and water (Yu et al., 1997). Gold plate (anode) and platinum plate (cathode) are used to produce gold nanorod particles with soaked silver plate under the ultrasonic vibration. The dip area of the silver plate can control an aspect ratio of the nanorod. The second one is called a seeding method, which uses the seed prepared by the chemical reduction for growing gold nanorod in the hydrogen tetrachloroaurate (III) tetrahydrate and cetyltrimethylammonium bromide (CTAB) (Murphy and Jana, 2002). The last one is photochemical method (Kim et al., 2002). In this method, gold ion is deoxidized with ultraviolet illumination of 254 nm. The aspect ratio is also controlled by the amount of silver ion in the solution.

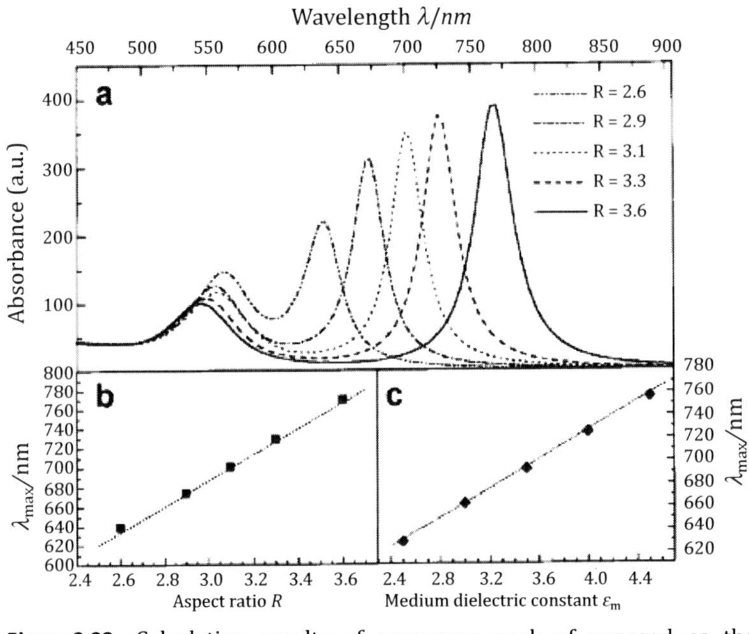

Figure 2.22 Calculation results of resonance peak of nanorod as the functions of (a, b) aspect ratio and (c) medium dielectric constant. Reprinted with permission from Stephan and Mostafa, 1999. Copyright 1999, American Chemical Society.

The establishment of these fabrication processes accelerated the study of gold nanorod. Its unique features are not only the wide range wavelength absorbance but also the high sensitivity of longitudinal plasmon mode toward environmental changes. It is pointed out that it has the potential to be adapted for new plasmonic devices.

2.3.2.2 Triangularly shaped nanoparticles by nanosphere lithography

To improve the uniformity of particle size and its arrangement, Van Duyne et al. have proposed a unique fabrication method for metal nanoparticles called nanosphere lithography (Jensen et al., 2000; Duval Malinsky et al., 2001; Haes and Van Duyne, 2002). For producing the metal nanoparticle, they prepared completely cleaned glass substrate as a first step. And then, they dropped polystyrene nanospheres on it. The nanospheres are self-assembled into a hexagonally close-packed 2D crystal structure that serves as a deposition mask. Silver was then deposited over the nanosphere array. After the depositions, the nanospheres are removed by sonicating the entire substrate in either CH_2Cl_2 or absolute ethanol for 2 min. An array of triangularly shaped nanoparticles remained on the glass substrate as a result. Figure 2.23 shows this fabrication procedure.

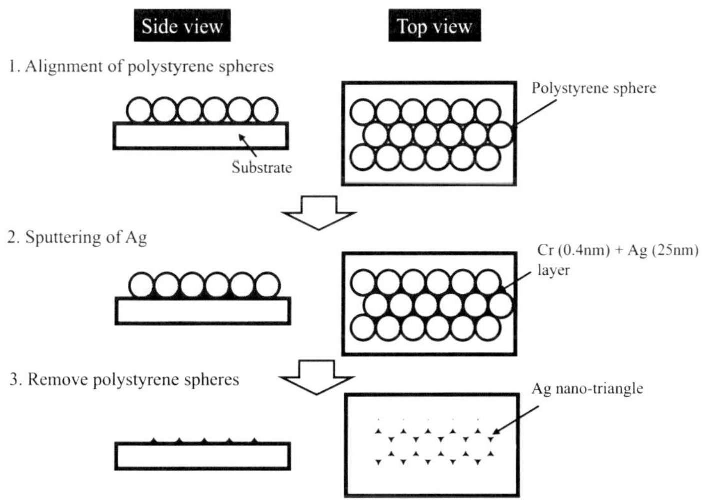

Figure 2.23 Schematic of fabrication process for nanoscale triangular particles on a substrate.

According to this fabrication process, chemical syntheses for metal colloids are not necessary and the controlled array of nanoparticles can be realized. Additionally, the optical response of the nanoparticles can be tuned up from UV to mid-near infrared wavelength regions by changing the thickness of silver and the diameter of polystyrene spheres. They proved that the absorbance peak wavelength is dependent on the refractive index of external environment. And the sensitivity, the wavelength shift value on the change of external refractive index, was 258 nm/RIU when SF-10 is used as a support substrate.

2.3.2.3 Closely arranged metal nanoparticles

When the metal nanoparticles are located away from each other, their resonance modes are not affected from neighbor nanoparticles and can be an independent mode. On the other hand, when they are placed within close distance, their resonant electric field reaches neighbor metal nanoparticles and it affects on the free electrons inside it. As a result, their resonance mode behaves differently from an individual mode.

Rechberger et al. (2003) have studied basic interactions between two cylindrical gold nanoparticles by observing their optical responses. They prepared particle pair samples with varying interparticle distance by electron beam lithography and observed a transmission spectrum. For a polarization direction parallel to the particle pair axis (p-mode), a red shift of the SPR peak wavelength was observed, whereas a blue shift was observed for a polarization direction perpendicular to the particle pair axis (s-mode). They also achieved to explain these behaviors by a simple dipolar interaction mechanism. On the resonant condition, the free electrons form a compression wave inside the metal nanoparticles. The neighboring resonant electric field works to avoid the recovery from the compressed distribution on the p-mode. It results in the lowering of the resonance frequency (red shift). On the s-mode, the resonance electric field works to accelerate the recovery of free electrons. And the resonance frequency increases (blue shift). Atay et al. (2004) have also studied congenial experiment and they examined the integrated metal particles in detail.

These results indicate that we need to concern about the interactions between the arrayed metal particles when they are closely arranged when designing the LSPR system.

2.4 Applications for Sensors

"Sensor" has been widely used for measuring the physical and chemical conditions of various materials and objects. Many kinds of sensing methods are proposed and applied as useful tools.

As presented above, the resonance condition of SPR sensitively depends on the dielectric constant near the metal surface. This characteristic has been studied for being applied to a sensor more than half a century ago. In 1983, Liedberg et al. have reported the first SPR biosensor adopting the antigen-antibody reaction principle (Liedberg et al., 1983). Pharmacia Biosensor corporation at that time (present GE Healthcare Corporation) has started to commercialize the biomolecular interaction analysis equipment (BIACORE: Biomolecular Interaction Analysis Core). The BIACORE series is widely used by the chemical, biological, and pharmaceutical researchers. In recent years, furthermore, new type of SPR sensor using LSPR which occurs on the nanosized metal particles has been proposed and studied in the nanophotonics field. In this section, we will introduce the principle and applications of these various SPR sensors.

2.4.1 Principle of SPR Sensors

For instance, the resonance condition of SPR under the Kretschmann configuration can be expressed by the following equation with reference to Section 2.2.2.

$$n \cdot \frac{\omega}{c} \cdot \sin\theta = \frac{\omega}{c}\sqrt{\frac{\varepsilon_{\mathrm{m}} \cdot \varepsilon_{\mathrm{s}}}{\varepsilon_{\mathrm{m}} + \varepsilon_{\mathrm{s}}}}$$

where n is the refractive index of prism. ω and c express the angular frequency and light speed, respectively. ε_{m} and ε_{s} are the permittivity of the metal and the neighboring material

This equation shows that θ depends on the ε_{s}, which is the dielectric constant of sample material. This means that the dark line angle (dip angle) of reflection light shifts when the sample constituent changes. To observe the reflection light at various angles, the light source is scanned or collimated by lens, generally. By detecting the dip angle, we can find the change of dielectric constant of sample material with high sensitivity (Fig. 2.24).

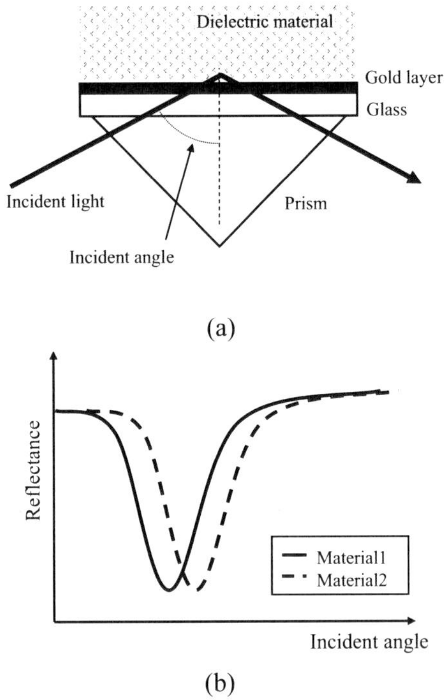

Figure 2.24 Schematics of (a) SPR sensor configuration and (b) detection principle.

Based on this principle, the SPR sensor was first proposed as gas sensor in 1982/1983 (Nylander et al., 1982/1983). In this paper, the sensitivity against the gas concentration was proved as high as 3×10^{-5} deg/ppm. As the next application, the SPR sensor is used for the liquid sensor that is popularly used recently. The features of this sensor are given as follows:

Advantages

- No pretreatment is necessary for the sample material.
- Real-time sensing can be realized.
- Very high sensitive detection especially near the sensor surface.

Disadvantages

- The signal drift will occur when the sample temperature changes.

- Calibration process is necessary for the absolute value detection.

2.4.2 SPR Biosensors

Biosensor is a tool to detect the biomolecules such as protein, DNA, and so on. This tool is widely used in the research field to analyze the biomolecular interactions and to find new medicinal biomolecules. This can also realize easy monitoring of health condition or rapid diagnosis of incipient disease by detecting the feature biomolecules included in the bodily fluids such as blood and urine. An easy-to-operate and higly sensitive biosensor is expected to be realized for improving our quality of life.

Various methods have been used to detect the biomolecules in years. The most popular way is to use the fluorescent molecules as the marker for target biomolecule. Though this method is widely used in the research and diagnosis field, a quantitative detection is still challenging due to the deactivation of fluorescent molecule. The difficulty of biosensor is that the size of target biomolecule is generally as small as about ten nanometers that is difficult to be observed by the general optical microscope. SPR biosensors utilizing the above features have been proposed and realized.

The principle of SPR biosensor is below. The probe molecules are immobilized on the sensor surface in advance. These probe molecules have the characteristic that they capture the complementary target molecules selectively. And it is well known that antibody, DNA, and peptide work as the probe molecules. Then the sample solution which may include the target molecules is sent over the sensor surface. When the target molecules are included in the sample solution, they are caught by the probe proteins. As the dielectric constant of biomolecule is commonly higher than that of buffer solution, the signal change (the shift of resonant condition) occurs (Jung et al., 1998). This detection method provides the quantitative measurement since the amount of signal shift depends on the density of captured target molecules. So, we can know the original concentration of the target molecules by detecting the amount of signal shift as shown in Fig. 2.25.

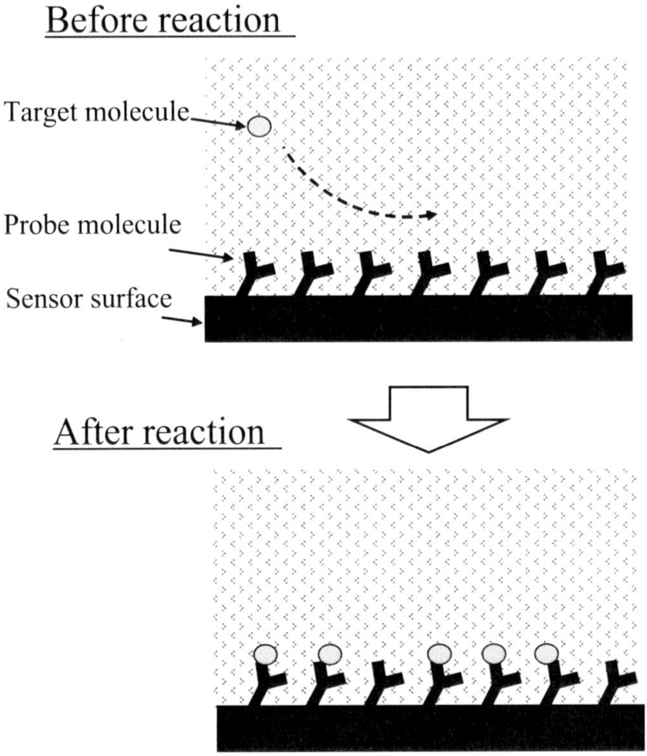

Figure 2.25 Reaction of target molecules on the SPR sensor surface.

Features of SPR biosensor are as follows:

(i) No pretreatment such as fluorescent modification is necessary for the sample solution.
(ii) Real-time sensing which provides much information about the target molecule such as association/dissociation constant can be operated.
(iii) Easy to use and quantitative detection can be realized.

For these useful features, SPR biosensor has been studied for use in a wide variety of field. It is being studied to be used for diagnosis, drug development, food analysis (allergen, toxic chemicals), environmental test, security, and so on. For realizing the ideal SPR biosensor, the technical issues must be addressed.

- The optical and photonic design of SPR devices.
- The immobilization of probe molecules and surface treatment.

- The fabrication of sensor substrate including the microchannel.
- The development of integrated sensing system.

As the practical application and research examples, following two development examples are introduced.

2.4.2.1 Biacore (GE Healthcare)

Biacore was founded under the name of Pharmacia Biosensor AB in Sweden in 1984. Its main product was the biomolecular interaction analyzing equipment based on SPR. In 1996, the company changed its name to Biacore AB Corporation. It was acquired by GE Healthcare in June 2006.

The company succeeded in the practical application of SPR to the first, and then, by continuing improvements, Biacore series are almost de facto standard in the current research field. The core technologies of Biacore equipments are as follows:

Optical and mechanical design On the Biacore equipment, SPR based on Kretschmann method is adopted as its detection principle. The sensor substrate is attached on the prism by using "optogel" without any optical gap in this system. This process is automated by mechanical mechanism. And this requires much less labor of the user than the conventional way of using matching oil between sensor substrate and prism. The incident light is focused by a lens and the light with multiple incident angles enters into the sensor surface on the condition of TIR. The reflection light is collimated by a lens and detected by CCD image sensor at once. The dip angle can be calculated by using approximation technique.

Microchannel Microchannel plays an important role for sending sample solution onto the sensor surface. The usability and accuracy of detection are directly affected by it. In Biacore system, microchannel made from silicon rubber in the system is pressed against the sensor substrate so that the sample solution should not leak from its clearance. By switching the valves over the microchannel mechanically, the flow path can be freely changed to fit the experiment.

Probe immobilization on surface of the sensor chip As described above, probe molecules need to be immobilized on the sensor surface before the interaction analysis. Many kinds of sensor surface on which probe immobilization layers such as dextran, amino group, and carboxyl group are prepared are on the market for Biacore series. And the standard protocols for the probe immobilization process are prepared in addition. User can immobilize the desirable probe molecules on it by using these products.

Analysis software More information can be got by SPR than the traditional detection methods as SPR is a real-time sensing method. To analyze the experimental data, dedicated analysis software is prepared. We can easily comprehend the affinity result by using this tool.

Recently, GE Healthcare focuses on the high-end model in Biacore series. And higher throughput and more automation mechanism are realized.

2.4.2.2 Optoquest Inc.

Optoquest Inc. is a company developing optical systems and optical devices in Japan. This company co-developed with our company and commercialized unique SPR equipment (compact SPR) in 2010 (Fig. 2.26). In compact SPR, the basic detection principle is adopting Kretschmann model as Biacore. The most characteristic point of this system is its sensor chip made from a plastic. The exterior appearance is shown in Fig. 2.27. The features are as follows:

Figure 2.26 Picture of commercialized compact SPR system.

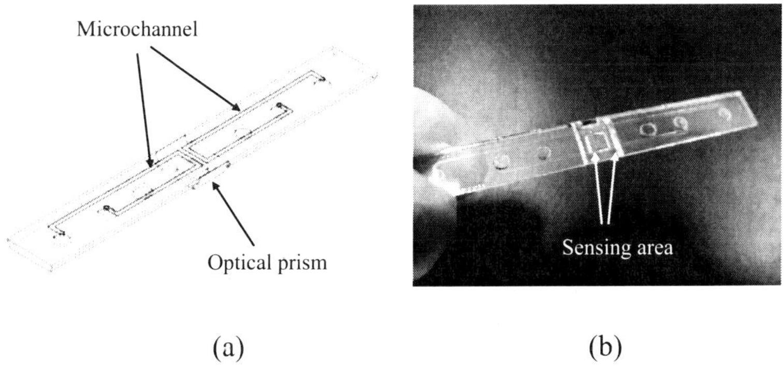

Figure 2.27 Integrated SPR sensor device: (a) CAD image and (b) picture.

Integrated prism The prism is combined in the sensor device as one device. User does not have to use the matching oil when setting the sensor chip. And no mechanical system is necessary for attaching on a prism with optogel. Moreover, measurement error caused by the air intervention between sensor and prism can be avoidable by this sensor device.

Integrated microchannel Two microchannels are also integrated on the surface of sensor chip. In the conventional system, the leakage of sample solution from the gap between sensor surface and microchannel can be problem. And frequent washings of microchannel are necessary to keep it clean. This device can reduce these problems by integrating microchannel and sensor device. The inlets and outlets of microchannels are connected at the bottom of sensor device with o-ring.

Film cover for microchannel The cover for microchannel consists of pressure adhesive film. As users easily conduct this process, they can select the immobilization of probe material using flow method or immersion method. Furthermore, as the film is transparent, observing the condition on the sensor during the experiment is possible. This feature leads to new potentials such as to be used with the fluorescent experiment.

The sensor device integrated with prism and microchannel can be fabricated by injection molding process at once. Therefore, low-cost device can be realized as the number of shipment increases. Since this device can be disposable, the usability is also much improved. This kind of new improvements is important for more prevalence of SPR equipments.

2.4.3 LSPR Biosensors

LSPR can also be used for biosensor as its resonance condition is sensitively changed by the dielectric constant of the ambient material. The features compared to the conventional SPR are as follows: LSPR does not require the TIR set up for its generation. As a result, the optical system can be simpler than the conventional SPR. The range of the resonant electric field corresponding to the sensing range can be smaller than the diffraction limit of light. The signal-to-noise ratio can be improved as the size of biomolecule is much smaller than the diffraction limit of light.

In recent years many studies about LSPR biosensors are addressed and reported (Sepúlveda et al., 2009). On the other hand, to generate LSPR, stable and high throughput fabrication technique of metal nanostructures is still a key issue. As an example, a study in which the metal nanostructure is fabricated by the nanosphere lithography method and a protein related with disease is introduced.

In 2004, Haes et al. proved that LSPR can be used for the disease-related protein detection. Using nanosphere lithography, they prepared triangular nanostructures whose perpendicular bisector is about 90 nm and height is 28–29 nm. In nanosphere lithography process, the polystyrene spheres (diameter is about 390 nm) are arranged on a substrate in close-packed structure. And then, Ag sputtering is conducted. The Ag particles come down through the clearance between spheres. And the arranged triangular nanostructures are formed. They immobilized ADDLs (amyloid-beta derived diffusible ligands) on them. ADDLs are present and significantly elevated in autopsied brain samples from humans with Alzheimer's disease. Then, varying the concentration of anti-ADDL IgG antibody and exposing the solution to the samples for 30 min completed the assay. They demonstrated that 10 nM concentration of the anti-ADDL antibody could be detected by their system. And they also found that non-specific binding occurred on the Cr layer, which is attaching layer between glass substrate and Ag nanoparticle. This work has shown that LSPR biosensor has the potential to be used for the clinical diagnostic for Alzheimer's disease. Continuous improvements have been conducted for LSPR biosensor. And the useful and small, low-cost biosensor is strongly expected to be realized by this method.

2.4.4 Waveguide SPR Sensors

Other than general SPR and LSPR sensor, the waveguide SPR has been developed. Figure 2.28 shows the basic structure of waveguide SPR device. A part of upper cladding is removed and thin gold layer is prepared on the core. The propagating light in the core interfere with the free electrons in gold layer at this part. On the resonant condition, the energy of the propagating light is transferred to the oscillation of free electrons. As the resonant condition is decided by the dielectric constant over the gold layer, this device can also work as a sensor. The features of waveguide SPR sensor are as follows:

(i) Remote sensing is possible by connecting the sensor device with the optical fiber.
(ii) The sensing space can be small because the detecting system can be separated from the sensing device.
(iii) Multiple detections are possible by fabricating fiber array configuration.

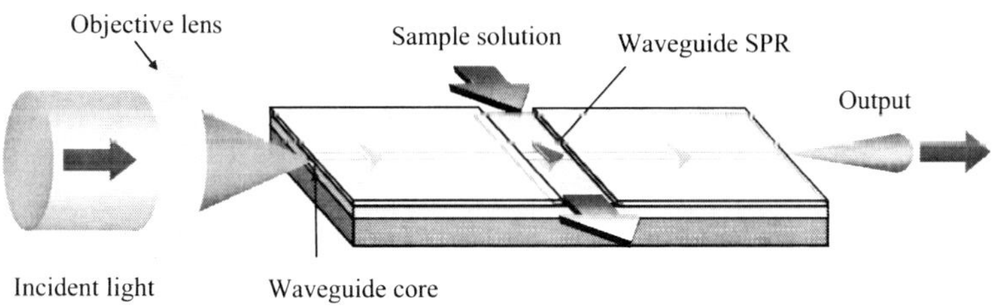

(a)

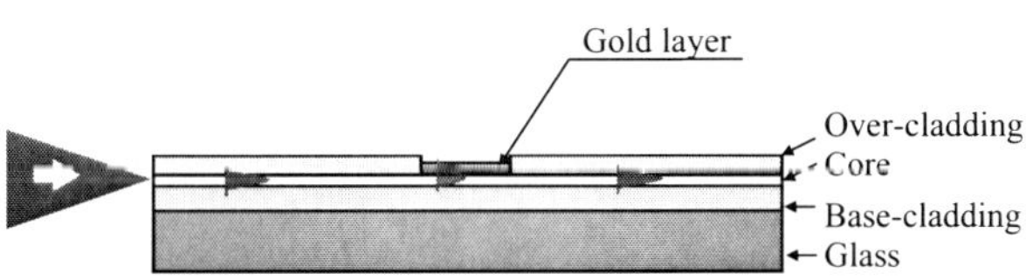

(b)

Figure 2.28 Schematics of waveguide SPR sensor structure: (a) birds-eye view and (b) cross-sectional image.

In the conventional method, the waveguide is fabricated through multiple semiconductor processes such as RIE (reactive ion etching) and so on. To reduce the process tact and cost, we have proposed a new fabrication process using imprint technology. Figure 2.29 shows the comparison of both processes. In our method, the core is prepared as groove by the first imprint process. And then the groove is filled with the polymer working as core. Finally, the upper cladding is produced by the cladding polymer except the sensing part and gold layer is sputtered. By fabricating the polymer, low-cost and easy tuning of the refractive index of core and cladding becomes possible.

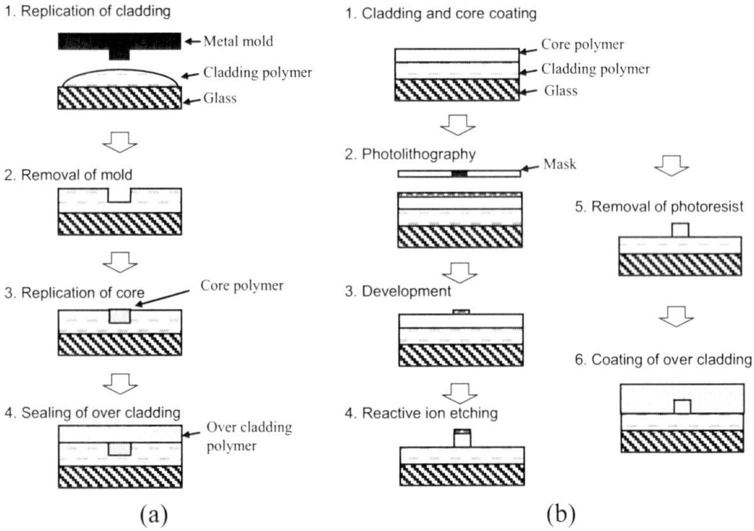

Figure 2.29 Fabrication processes of waveguide device: (a) based on the imprint method and (b) a conventional process using semiconductor fabrication process.

Figure 2.30 shows the fabricated sample of the waveguide SPR device. The width and height of core are both 3 μm for demonstrating as single mode waveguide. Along the core, a part of cladding about 2 mm length is removed and the gold layer (thickness, 53 nm) is formed. In our experiment, the adhesion layer between core polymer and gold was not necessary. On the top of the gold layer, TiO_2 layer (thickness, 15 nm) is prepared as the buffer layer for tuning the sensing range of the refractive index.

To demonstrate that this can work as a sensor, the different ratio mixtures of purified water (n = 1.33) and ethylene glycol (n = 1.43) are detected by this device. The experimental set-up is shown in Fig. 2.31. The laser (wavelength, 632 nm) was collimated by an object lens and focused at the end face of core. The *x*, *y*, and *z* position of waveguide is adjusted by using stepping motor. The output light from the other end surface is split to CCD for detecting the coupling condition of incident light and power meter for measuring the power of the guided light.

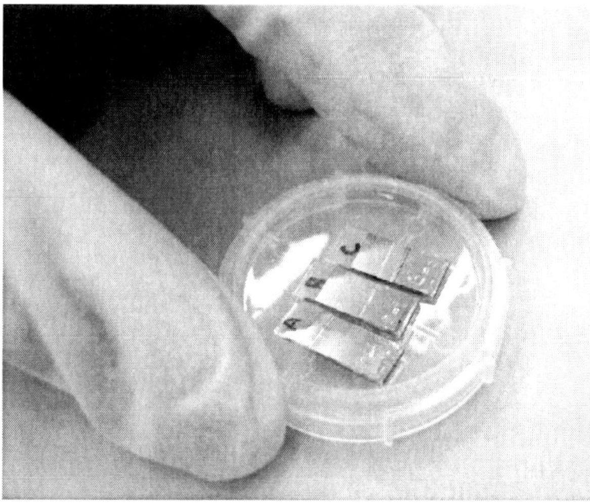

Figure 2.30 Picture of the fabricated waveguide SPR devices based on the imprint technique.

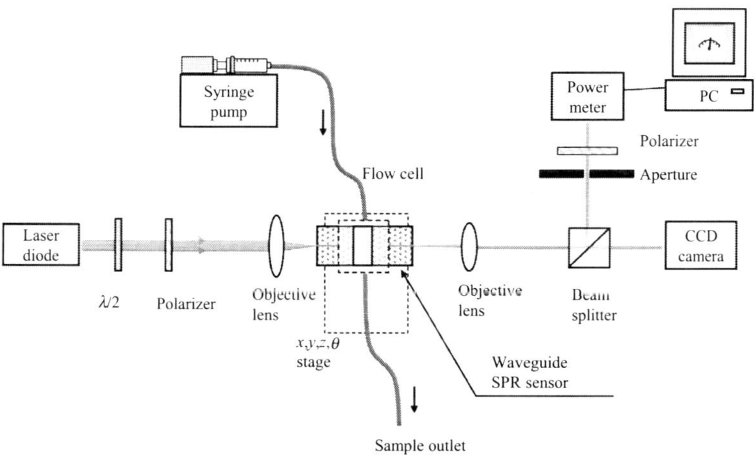

Figure 2.31 Schematic of experimental set-up for waveguide SPR sensor.

The result shows that the strength of the TM component of the guided light changes according to the refractive index of sample solution (Fig. 2.32). This means that this device can work as a sensor. And the TE component of the guided light does not change since it does not generate SPR in the waveguide. We, moreover, compared this result with the simulation result (Fig. 2.33). The simulation program was home-made one based on TMM. The result had a good agreement with the experimental result. And this confirms the experimental result is reasonable and proper. Waveguide SPR has the unique features described above and can be used as an integrated sensor in the future.

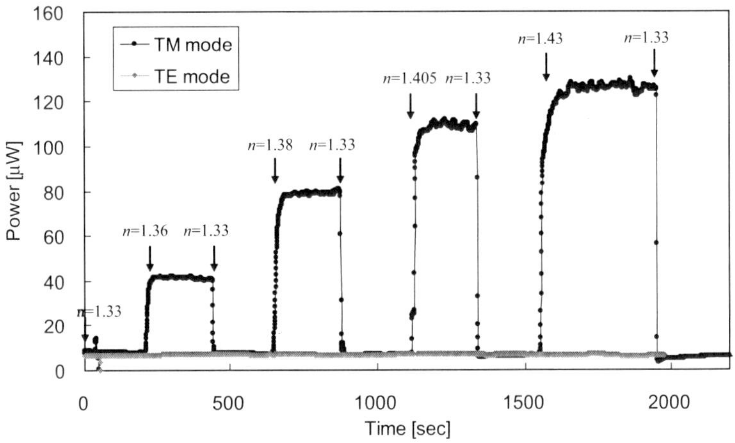

Figure 2.32 Experimental result of waveguide SPR sensor in that the refractive indexes of the sample material are changed from 1.33 to 1.43.

2.5 Other Applications

SPR has the characteristics that strongly intensified electric and magnetic fields are generated and confined in less than a few hundred nanometers. Making use of these features, new applications are proposed and studied:

- High density data storage
- High efficiency thin-film photovoltaic device

- Nanophoto diode
- Plasmon laser
- Organic electro luminescence
- Plasmonic LED device
- Color filtering device
- Drug delivery
- SERS (surface-enhanced Raman scattering) device
- Negative refractive index, metamaterial

In this section, some typical examples are introduced.

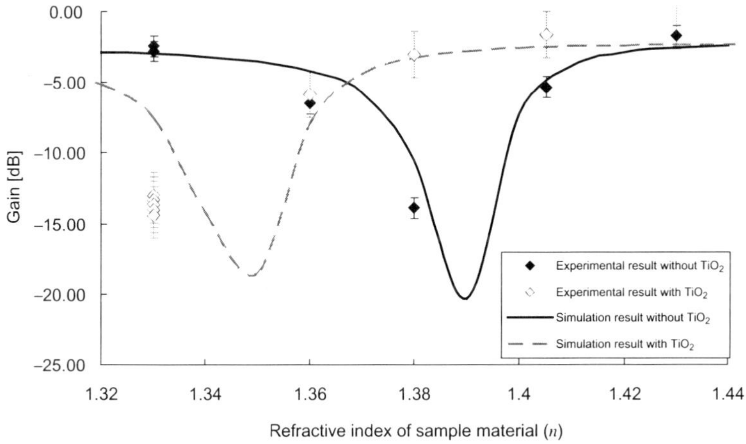

Figure 2.33 Output/input gain results of experiment and simulation with/without TiO_2 layer (thickness, 15 nm).

2.5.1 Photovoltaic Device

Photovoltaic device has been expected as one of the main renewable energy resources. In the solar cells made of highly purified materials such as single crystal silicon and compound semiconductor, high conversion efficiency close to the theoretical limit has been already achieved. Hence, the issue for these types of solar cells is becoming a reduction of material cost. On the other hand, in the thin film solar cell of organic-based systems and amorphous silicon, conversion efficiency and extension of life span are still challenging while the material cost is lower than that of crystal type and the fabrication process is also simple. To solve these problems, the applications of innovative methods such as SPR have been addressed as there is a limit only by the conventional attempts.

By using SPR and LSPR, highly intensified electric field can be confined in a very close region to the metal surface. This feature is expected to contribute to improving the power conversion efficiency and to thinning the thickness of solar cell material.

The attempt to apply the plasmon phenomena for photovoltaic device has started from about 1995. Typical configurations proposed so far can be classified into following three methods (Fig. 2.34).

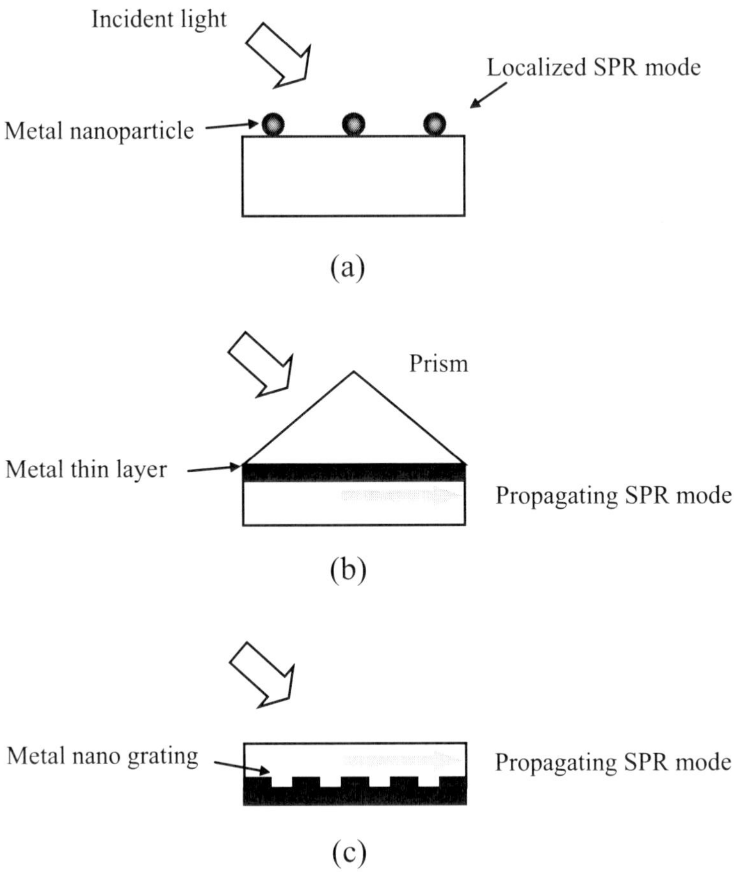

Figure 2.34 Configurations of SPR generation for photovoltaic devices.

2.5.1.1 Use of LSPR

The method shown in Fig. 2.34(a) makes use of the intensified local resonant electric field near the metal nanostructures such as metal nanoparticles and metal nano-islands. Just by preparing metal nanoparticles on the photovoltaic material, this system can be realized. And various ways based on this method are proposed for its easiness. The typical one is to capture the exposed light by metal nanoparticles on a device surface as LSPR. And then, the enhanced scattering light from the metal nanoparticle is absorbed.

Especially in the organic thin film semiconductor, the diffusion length of excitons is so short that only the excitons generated in the very vicinity of the joint surface of donor and acceptor can be desterilized. In this case, by arranging the metal nanoparticles on the joint surface, the conversion efficiency is expected to be much improved due to the locally enhanced electric field. Many studies based on this principle are reported in recent years. As the width of absorption spectra of general LSPR is small, the absorption spectra should be, however, broadened for more efficient use of solar power.

2.5.1.2 Use of attenuated total reflectance type of SPR

In the propagating SPR, there exist many modes in wide wavelength range and the penetration depth of resonant electric field is a few hundred nanometers which is much deeper than that of LSPR. Therefore, the use of SPR is also expected for improving the conversion efficiency of photovoltaic device.

Figure 2.34(b) shows the ATR configuration for generating SPR. The idea to use ATR SPR for improvement of photovoltaic conversion efficiency is proposed by Hayashi et al. in 1993 (Kume et al., 1993). And they proved the conversion efficiency can be improved 2.3 times higher than one without SPR on the condition that the incident light enters into organic solar cell-based CuPc at the incident angle of 45 degree (Kume et al., 1995). In this method, however, only light entering at a certain angle can be efficiently used for the energy conversion. For practical application, some improvements must be necessary.

2.5.1.3 Use of grating type of SPR

In Fig. 2.34(c), the periodic grating is used for generating SPR. Ishi et al. (2005) reported that the grating SPR can be used for improving the efficiency of Si photodiode in 2005. This technique was applied and demonstrated that it can improve the conversion efficiency of organic thin film solar cells 3.2 times higher than the conventional one (Lindquist et al., 2008). It is, however, pointed out that periodic grating limits the range of resonant wavelength. And some challenges exist in this method, too.

In Chapter 4, we have demonstrated that the range and width resonant wavelength can be tuned up by using the nanoimprint LSPR method. So, we believe that it has the potential to be used for the improvement of photovoltaic energy conversion efficiency, too.

2.5.2 Nano-photodiode

Photodiode plays an important role as a signal receiving device for sensors and information transmission. Since optical communication technology has the strong advantages such as a large capacity of the transmission information, electromagnetic noise immunity, and radiation improvement, it progresses and spreads rapidly. As a result, the improvement of sensitivity and response speed of photodiode are demanded. Here, recent activities about them will be introduced.

The major factor to limit the response speed of photodiode for optical signal is two-fold. One is photon transport time between the positive and negative electrodes of the semiconductor surface. Another one is the junction capacitance of the photodiode.

As to deal with the former, there may be mentioned to shorten the distance between the electrodes. There is, however, a problem that it will tend to slow down the speed of light absorption resulting from the reduced phonon assist especially in a silicon case. For the latter, though it is necessary to use a lens to focus the incident light into small area for reducing the junction area, the conventional lens cannot focus the light into a smaller area than the diffraction limit of light and is not easy to be prepared on the substrate by a semiconductor process all at once.

Thio et al. (2001) have proposed a plasmon antenna with the metal grating structure of concentric rings known as the Bull's eye

(Lezec et al., 2002). Period of the grating is corresponding to the wavelength of the surface plasmon. The light irradiated from the top couples to the SPR by metal grating and the generated surface plasmon propagates to the perforated hole located at the center of concentric rings. As a result, the strength of the near-field light at the center hole can be extraordinarily high. The metal grating can serve as the role of the anode as it can be made of metal. By reducing the semiconductor active region limited to the vicinity of the central hole, it is possible to significantly reduce the junction capacitance. In addition, due to the enhancement of electric field by the resonance, a sufficient response speed can be obtained even under the condition that the active junction area is small. It is demonstrated that FWHM (full width at half maximum) of the response to high-speed optical pulse to be as small as 22 ps when the bias voltage is 0 V (Fujikata et al., 2005).

By using SPR to the photodiode in this way, we can produce benefits such as improved speed of response. On the other hand, the drawbacks such as loss due to scattering caused by SPR need to be solved, hereafter.

2.5.3 Color Filter

Color filter used in the liquid crystal displays and image processing apparatus is one of the important components that are indispensable to the advanced information society. Conventional color filter is dependent on the optical properties of the pigments, and the development of the pigment with excellent transmission characteristics is an important issue. In addition, the color filter is configured with a black matrix resist for preventing the mix of the dispersed pigment with other colors, which is also an issue of process complexity and loss of transmitted light. If free control of light transmission characteristics and a new fabrication process of color filter based on batch process are realized, the performance of the color filter can be dramatically improved.

Metal thin film with a periodic nanohole array has been the focus of attention since it was found to transmit a specific range of the wavelength of incident light extraordinarily (Ebbesen et al., 1998). Its principle is assumed that the SPR is generated on the periodic nanohole array according to the grating effect and that the generated SPR goes through the nanohole and couples to light again.

In 2003, it is demonstrated that the transmitted light by a nanohole array fabricated by FIB (focused ion beam) can be corresponding to the three primary colors of RGB as a first example of SPR color filter (Barnes et al., 2003).

The basic transmission characteristic of SPR color filter can be adjusted by changing the period and diameter of nanohole array, which means it has the potential to achieve high-performance filter characteristics unprecedented. On the other hand, since the present process throughput of nanohole array by an electron beam method is too low to be used as a practical device, improvement of fabrication process such as the use of nanoimprint method is necessary.

Chapter 3

Nanobiology

3.1 Design of Immobilization Layer for a Surface Plasmon Resonance Biosensor

For surface plasmon resonance (SPR) biosensing in solution, it is necessary that one interaction partner (probe molecules) is immobilized onto the sensor surface to capture the target molecules. Conventional immunochemical methods such as enzyme-linked immunosorbent assay (ELISA) are based on the simple physical adsorption of probe molecules onto a plastic plate. However, it is thought that a more sophisticated approach is required for surface preparation of a sensor surface for SPR biosensing. This is because the sensitivity of SPR biosensors is highly dependent on the binding capacity of the immobilized probe molecules on the sensor surface and on the resistance of the surface to nonspecific protein adsorption. The performance of the sensor surface is supported by three crucial factors (the capture agent, the surface chemistry, and the surface matrix). Figure 3.1 shows a schematic diagram of an immobilization layer for SPR biosensing and the desired characteristics of the three key factors.

3.1.1 Key Factor 1: Capture Agent

In principle, it is difficult for SPR sensors to clearly distinguish the signal component of target molecules from the background noise

Nanoimprint Biosensors: The Fusion of Nanofabrication, Nanophotonics, and Nanobiology
Takeo Nishikawa and Satoshi Fujita

ISBN 978-981-4316-67-5 (Hardcover), 978-981-4463-42-3 (eBook)
www.panstanford.com

factors associated with nonspecific absorption. First, the capture agent must have the capability for specific recognition of the target molecules. Proteins such as immunoglobulin G (IgG), which are known as immune antibodies, are frequently used as capture agents on account of their high specificity toward their target antigens (Table 3.1; Besselink et al., 2004; Yang et al., 2005). Second, the selection of capture agents with high affinity (equilibrium dissociation constant, in units of molar concentration; $K_D < 10^{-9}$) is necessary to achieve high sensitivity. Recently, nucleic acid aptamers and synthetic peptides have been developed as artificial antibodies with high specificity, high affinity, and ease of size control, using the molecular evolutionary systematic evolution of ligands by exponential enrichment (SELEX) process and phage display method. The obtained artificial antibodies have often been used as capture agents in SPR (Katz et al., 1995; Polonschii et al., 2010).

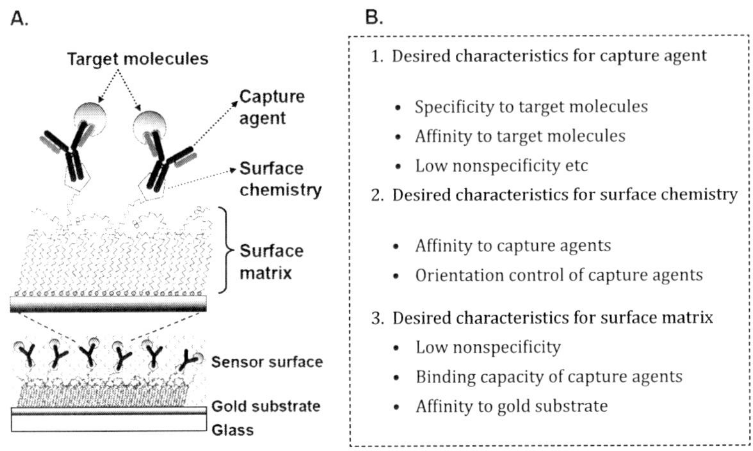

Figure 3.1 (A) Schematic diagram of the immobilization layer and (B) desired characteristics for the three key factors.

3.1.2 Key Factor 2: Surface Chemistry

A coupling method involving activated *N*-hydroxysuccinimide (NHS) esters is one of the most commonly used surface chemistry techniques for anchoring capture molecules to a sensor surface (Table 3.1; Johnsson et al., 1991; Lahiri et al., 1999). Because the target of this activated ester is any amino group that is present on

Table 3.1 Examples of three factors (capture agent, surface chemistry, and surface matrix) that are important in the formation of immobilization layers

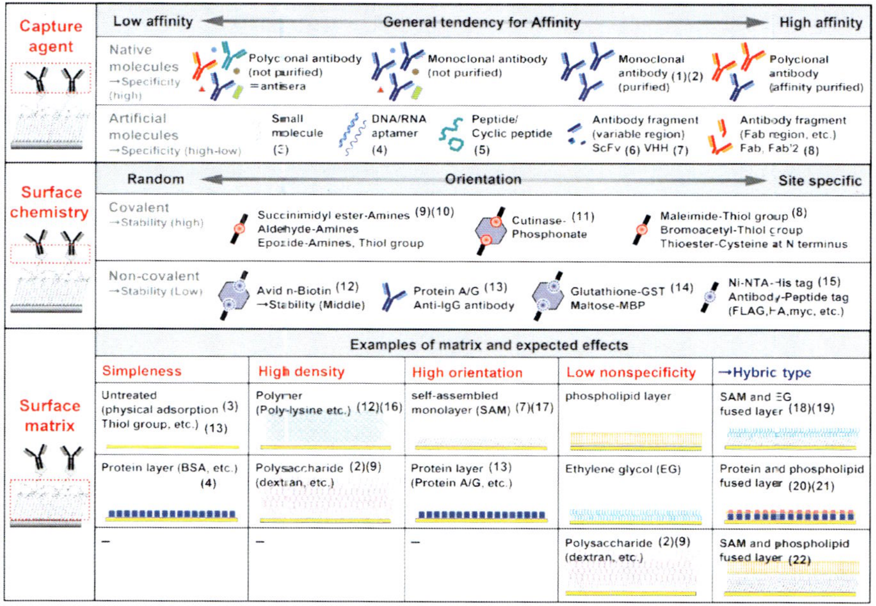

Capture agent	Low affinity ← General tendency for Affinity → High affinity			
Native molecules →Specificity (high)	Polyclonal antibody (not purified) =antisera	Monoclonal antibody (not purified)	Monoclonal antibody (1)(2) (purified)	Polyclonal antibody (affinity purified)
Artificial molecules →Specificity (high-low)	Small molecule (3); DNA/RNA aptamer (4)	Peptide/ Cyclic peptide (5)	Antibody fragment (variable region) ScFv (6) VHH (7)	Antibody fragment (Fab region, etc.) Fab, Fab'2 (8)
Surface chemistry	Random ← Orientation → Site specific			
Covalent →Stability (high)	Succinimidyl ester-Amines (9)(10) Aldehyde-Amines Epoxide-Amines, Thiol group	Cutinase- (11) Phosphonate	Maleimide-Thiol group (8) Bromoacetyl-Thiol group Thioester-Cysteine at N terminus	
Non-covalent →Stability (Low)	Avidin-Biotin (12) →Stability (Middle)	Protein A/G (13) Anti-IgG antibody	Glutathione-GST (14) Maltose-MBP	Ni-NTA-His tag (15) Antibody-Peptide tag (FLAG,HA,myc, etc.)

Surface matrix	Examples of matrix and expected effects				
	Simpleness	High density	High orientation	Low nonspecificity	→Hybric type
	Untreated (physical adsorption (3) Thiol group, etc.) (13)	Polymer (Poly-lysine etc.) (12)(16)	self-assembled monolayer (SAM) (7)(17)	phospholipid layer	SAM and EG fused layer (18)(19)
	Protein layer (BSA, etc.) (4)	Polysaccharide (2)(9) (dextran, etc.)	Protein layer (13) (Protein A/G, etc.)	Ethylene glycol (EG)	Protein and phospholipid fused layer (20)(21)
	–	–	–	Polysaccharide (2)(9) (dextran, etc.)	SAM and phospholipid fused layer (22)

(1) Besselink et al., 2004; (2) Yang et al., 2005; (3) Kim et al., 2007; (4) Polonschii et al., 2010; (5) Katz et al., 1995; (6) Torrance et al., 2006; (7) Huang et al., 2005; (8) Bonroy et al., 2006; (9) Johnsson et al., 1991; (10) Lahiri et al, 1999; (11) Kwon et al., 2004; (12) Huang et al., 2002; (13) Lee et al., 2007; (14) Ha et al., 2007; (15) Sigal et al., 1996; (16) Wazawa et al, 2006; (17) Masson et al., 2007; (18) Prime and Whitesides, 1993; (19) Sigal et al., 1998; (20) Athey et al., 2005; (21) Shah et al., 2007; (22) Ishizuka-Katsura et al., 2008.

the protein molecule with high probability, this coupling method is applicable to various capture agents used for SPR biosensing. To obtain highly sensitive SPR signals, the orientation of capture agents should be considered. The percentage of biochemically active capture agents that can interact with the target molecules would be higher if the orientation of the capture agents on the sensor surface can be improved. As a result of this improvement, the SPR response would be increased several times. As a typical example, the surface chemistry to immobilize proteins via hexahistidine tags (His-tag) has been used (Sigal et al., 1996). In recent years, mutated proteins, such as functional fusion proteins, have been used for achieving oriented immobilization of capture agents and for simplification of the immobilization process (Terrettaz et al., 2002; Ha et al., 2007; Park et al., 2009; Le Brun et al., 2011). Some mutated proteins are already on the market as commercial layers (Athey et al., 2005).

3.1.3 Key Factor 3: Surface Matrix

Polymers, polysaccharides, self-assembled monolayers (SAMs), and phospholipid and protein layers, among other, have all been reported as surface matrices (Table 3.1). One of the most important functions of the surface matrix in SPR biosensors is the suppression of nonspecific adsorption of contaminants to the sensor surface. For this purpose, the introduction of oligo(ethylene glycol) (OEG) molecules is highly effective (Prime and Whitesides, 1993; Sigal et al., 1998). Moreover, it is also important to increase the binding capacity of the capture agents. This factor, which determines the maximum signal variation of the SPR sensor, can control the dynamic range of biosensing. One example of a surface matrix that has been successful in increasing the binding capacity of capture agents is the carboxymethylated dextran matrix provided by GE Healthcare (Sweden). The carboxymethylated dextran matrix provides a three-dimensional space with a thickness of 100 nm for target molecule binding (Johnsson et al., 1991; Yang et al., 2005).

3.2 Improved Antibody Immobilization on an OEG-Mixed SAM

An SPR sensor chip with a carboxymethyl dextran (CM5) system comprising a layer of carboxymethyl dextran on the sensor surface has been commercialized by GE Healthcare (Löfås et al., 2006). For immobilization of antibodies, the IgG protein is allowed to adsorb to the negatively charged CM5 system at an acidic pH by an anion-exchange effect, and then the protein is conjugated to the carboxyl groups of the carboxymethyl dextran using 1-ethyl-3-(3-dimethylaminopropyl)carbodiimide hydrochloride (EDC) and NHS (Löfås et al., 2006). The CM5 chip has been widely used (Yang et al., 2005; Skottrup et al., 2007; Stapleton et al., 2009; Crosson et al., 2010; Nilsson et al., 2010) because it exhibits not only a high capacity for antibody immobilization, but also a wide dynamic range for protein detection. However, the CM5 chip may suffer from nonspecific adsorption of proteins at high concentration and steric hindrance of antigen binding due to overcrowding between the immobilized antibodies (Yang et al., 2005; Situ et al., 2008).

In contrast, SAMs on planar gold surfaces have also been widely used for immobilizing proteins such as antibodies on the SPR sensor chip surface (Lahiri et al., 1999). In this configuration, the SAMs containing OEG have been known to show resistance to nonspecific protein adsorption (Kim et al., 2007; Kawaguchi et al., 2007; Nagatomo et al., 2009); moreover, mixed SAMs composed of carboxyl group-terminated and hydroxyl group-terminated OEG alkanethiols have been shown to be very effective in reducing nonspecific protein adsorption presumably due to electrostatic attraction of opposite charged proteins (Pale-Grosdemange et al., 1991; Prime and Whitesides, 1991, 1993; Kyo et al., 2005). Therefore, such OEG-mixed SAMs would be more advantageous than the CM5 chip in some circumstances such as the detection of biomarkers in blood. In addition, a recent study demonstrated that OEG-mixed SAMs were resistant to air or N_2 and could retain biosensor capability for more than 30 days (Jans et al., 2008). Thus, a SAM mixture comprising carboxy-terminated and hydroxyl-terminated OEG-mixed SAM is likely to be very suitable for use as a protein-immobilization system for SPR-based biosensors.

In this study, we investigated the optimal conditions for antibody immobilization on the carboxy-terminated and hydroxyl-terminated OEG-mixed SAMs for highly sensitive SPR biosensing. Alpha-fetoprotein (AFP), a 70 kDa, well-known biomarker for hepatocellular tumors, was chosen for use as a standard antigen in the present SPR immunosensing study. An anti-AFP antibody was immobilized on the mixed SAM through EDC/NHS coupling chemistry, and AFP was detected by SPR. The SPR detection of AFP was conducted using aqueous solutions containing AFP and fetal bovine serum (FBS) to emulate AFP in blood. The optimal condition was thus investigated to accomplish the highest signal-to-noise ratio for AFP detection.

3.2.1 Materials and Methods

3.2.1.1 Materials

The following materials were purchased from Dojindo Laboratories (Kumamoto, Japan):

1. Carboxy-EG6-undecanethiol
2. 20-(11-Mercaptoundecanyloxy)-3,6,9,12,15,18-hexaoxaeicosanoic acid, ($HS(CH_2)_{11}(OCH_2$-$CH_2)_6OCH_2CO_2H$), (OEG-COOH)
3. Hydroxy-EG3-undecanethiol
4. 11-mercaptoundecanol triethyleneglycol ether, $HS(CH_2)_{11}(OCH_2CH_2)_3OH$) (OEG-OH)

Bare gold substrates (SIA-kit), Sensor Chip CM5 (research grade), amine coupling reagents containing EDC, NHS, ethanolamine hydrochloride (pH 8.5), HEPES-buffered saline (HBS) buffer (pH 7.4) containing 10 mM *N*-(2-hydroxyethyl)piperazine-*N′*-2-ethanesulfonic acid (HEPES), 0.15 M sodium chloride, 3 mM ethylenediaminetetraacetate (EDTA), and 0.005% (v/v) Surfactant P20 were obtained from GE Healthcare (Uppsala, Sweden). An affinity purified antibody (95% Rabbit IgG) against human AFP was purchased from Monosan (Uden, The Netherlands). Pure human AFP (a single band on SDS-PAGE) was obtained from Morinaga Institute of Biological Science (Yokohama, Japan). Undiluted FBS was purchased from Japan Bioserum (Hiroshima, Japan). All other chemicals were obtained from Wako Pure Chemical Industries (Osaka, Japan). Ultrapure water prepared by a Milli-Q Synthesis A10 system (Millipore, Billerica, MA, USA) was used for preparation of all solutions.

3.2.1.2 Preparation of mixed SAM surface

SIA-kit substrates were incubated in a piranha solution (96.0–98.0% sulfuric acid: 30.0–35.5% hydrogen peroxide = 70:30, v/v) for 30 min at room temperature to remove all organic contaminants on the surface of the gold substrates. After thorough rinsing with ultrapure water, the gold substrates were immediately immersed in an ethanol solution containing 1 mM OEG-OH and 1 mM OEG-COOH. After 24 h of SAM deposition, the substrates were rinsed with ethanol, dried under a stream of nitrogen, and then stored at 4°C in nitrogen until use.

3.2.1.3 Preparation and evaluation of antibody-immobilized surface

An SPR Biacore 2000 in combination with Biacore 2000 Control Software (Version, 3.2.1) (GE Healthcare, Uppsala, Sweden) was used to monitor reactions taking place on the OEG-mixed SAM modified sensor surface coupled with the flow cell of Biacore 2000 at 25.0°C.

The immobilization of anti-AFP on the OEG-mixed SAM (see above) was carried out by using carbodiimide coupling chemistry with EDC and NHS (Lahiri et al., 1999). The carboxylic groups on the SAM surface were activated for 10 min by injecting an aqueous solution containing 0.2 M EDC and 50 mM NHS at a flow rate of 5 μL/min. Subsequently, an antibody solution containing 20 μg/mL anti-AFP and 10 mM sodium acetate buffer (pH 5.5) was injected for an appropriate time (see Results and Discussion) at a flow rate of 5 μL/min for the coupling of anti-AFP to the OEG-mixed SAM. After the coupling reaction, an aqueous solution containing 1 M ethanolamine at a pH of 8.5 was injected for 10 min to block the remaining NHS ester groups on the OEG-mixed SAM (Fig. 3.2A,B). Furthermore, the antibody-immobilized surface was washed for 1 min with a 50 mM sodium hydrochloride (NaOH) solution to remove the antibodies nonspecifically adsorbed on the sensor surface, and then, thoroughly rinsed with HBS buffer.

To measure the interaction of AFP with the anti-AFP immobilized on the SPR sensor chip, 1 μg/mL AFP dissolved in HBS buffer was injected at a flow rate of 20 μL/min for 2.5 min. The AFP concentration was determined using the Abbott AFP-EIA kit according to the manufacturer's instructions. After surface regeneration with 50 mM NaOH for 1 min, the undiluted FBS was injected at a flow rate of 20 μL/min for 2.5 min to measure nonspecific protein adsorption

(Fig. 3.2C). The SPR response in resonance units (RU) is defined as 1 RU = 0.0001° of resonance angle shift (Jönsson et al., 1991; Lahiri et al., 1999).

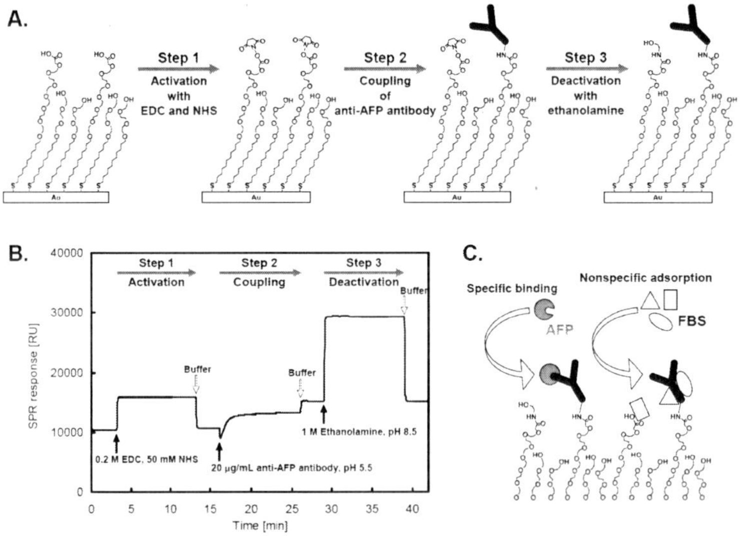

Figure 3.2 Schematic representation of antibody-immobilization on the mixed self-assembled monolayer (SAM) sensor and assay for performance of the sensor. (A) Stepwise processes of the antibody-immobilization on the oligo(ethylene glycol) (OEG)-mixed SAM surface are illustrated: (Step 1) activation of the carboxylic groups of OEG-COOH on the OEG-mixed SAM surface using 0.2 M 1-ethyl-3-(3-dimethylaminopropyl)carbodiimide hydrochloride (EDC) and 50 mM *N*-hydroxysuccinimide (NHS) to form NHS ester groups; (Step 2) coupling reaction between the NHS ester-activated SAMs and anti-alpha-fetoprotein (anti-AFP) antibodies by displacement of the NHS ester groups by the primary amino groups of the antibody; and (Step 3) deactivation of the remaining activated NHS ester groups using 1 M ethanolamine (pH 5.5). (B) A typical surface plasmon resonance (SPR) sensorgram in the antibody-immobilization processes using 25% OEG-COOH-mixed SAM. After the each step, the surface was immediately rinsed with HEPES-buffered saline buffer (pH 7.4) containing 10 mM HEPES, 0.15 M sodium chloride, 3 mM EDTA, and 0.005% (v/v) Surfactant P20. (C) An assay for performance of the antibody-immobilized mixed SAMs shown in Figs. 3.3 and 3.4. As indicators of specific binding and nonspecific adsorption, AFP and undiluted fetal bovine serum (FBS) were used, respectively.

3.2.1.4 Preparation of antibody-immobilized CM5 surface

Carboxylic groups on the CM5 chip were activated using an aqueous solution containing 0.2 M EDC and 50 mM NHS, the coupling reaction was conducted with a 20 μg/mL anti-AFP solution in a 10 mM sodium acetate buffer at a pH of 5.5, and then, the deactivation of the remaining NHS ester groups was made using a solution containing 1 M ethanolamine at pH 8.5 according to the manufacturer's instructions (Johnsson et al., 1991).

3.2.2 Optimization of Mixing Condition of the Mixed SAMs

It has been reported that IgG antibodies accumulate on the OEG-mixed SAM at an acidic pH owing to electrostatic interaction, so-called pre-concentration (Johnsson et al., 1991), and then effectively immobilized on the mixed SAM by EDC/NHS chemistry (Lahiri et al., 1999). Therefore, the coupling reaction of anti-AFP was carried out in a solution at pH 5.5.

However, the negative charge on the OEG-mixed SAM may often induce nonspecific protein adsorption at neutral pH during antigen–antibody reactions. This type of protein adsorption can be reduced if the molar fraction of OEG-COOH in the mixed SAM is decreased. For this purpose, OEG-mixed SAMs with various percentages (expressed as mol%) of OEG-COOH were prepared. Anti-AFP was allowed to react with the activated NHS ester on the OEG-mixed SAM for 10 min. After the cross-coupling reaction of anti-AFP on the mixed SAM, the amount of anti-AFP immobilization was evaluated. On the 100% OEG-COOH SAM, the SPR signal increment due to antibody immobilization was 2745 RU (Fig. 3.3B). According to Stenberg et al. (1991), an SPR signal of 1000 RU corresponds to 1 ng/mm^2 of immobilized protein. However, in using this correlation coefficient, the configuration of the immobilized antibody molecules on the gold substrate should be considered. In this respect, the molecular number of antibodies bound on the 100% OEG-COOH SAM is computed to be about 1.1 × 1010 molecules/mm^2. The theoretical maximum binding of IgG on a flat surface is estimated to be about 0.71 × 1010 molecules/mm^2 (assuming a prone orientation) or

2.0 × 1010 molecules/mm^2 (assuming a standing orientation) using X-ray crystallography data of IgG (10 × 14 × 5 nm) (Amit et al., 1986). Hence, the antibodies bound on the 100% OEG-COOH SAM are expected to be in a prone orientation in high density, or at least in a partly standing orientation. Importantly, the highest amount of anti-AFP immobilization at 4516 RU was achieved on the 25% OEG-COOH-mixed SAM (OEG-COOH:OEG-OH = 25:75, mol%), and the molecular number of antibodies was calculated to be about 1.8 × 1010 molecules/mm^2. Therefore, those antibodies would be in a close-packed standing orientation.

After anti-AFP immobilization, the SPR signal intensity increase was measured by injecting 1 μg/mL AFP solution (Fig. 3.3C). The maximum amount of AFP binding (265 RU) was achieved on the 25% OEG-COOH-mixed SAM, and this signal increase corresponded to a molar antigen-to-antibody ratio of 0.126. The result indicates that 12.6% of the anti-AFP molecules immobilized on the 25% OEG-COOH-mixed SAMs took part in AFP binding.

Furthermore, nonspecific adsorption on the mixed SAM was examined using undiluted FBS, which contained a total of 35.6 mg/mL of serum proteins according to the manufacturer's instruction. The nonspecific adsorption was significantly high on the 100% OEG-COOH SAM (Fig. 3.3C). This may be mainly due to electrostatic interaction between the serum proteins and the ionized carboxyl groups on the OEG-COOH SAM. On the mixed SAMs of 50% and 25%, the nonspecific adsorption was significantly lower (Fig. 3.3C). However, the nonspecific adsorption was observed to be slightly higher on the 10% OEG-COOH SAM and significantly higher on the 5% OEG-COOH SAM. The results suggest that the amount of nonspecific adsorption is influenced by the degree of immobilization of anti-AFP on the mixed SAMs in addition to the electrostatic interaction between the ionized carboxyl groups on the OEG-COOH SAM and the serum proteins.

Thus, as a result of the present optimization study, nonspecific protein adsorption was minimized on the 25% OEG-COOH-mixed SAM, as determined from the signal-to-noise ratio, and the AFP binding relative to the nonspecific adsorption of FBS showed a maximum value in the mixed SAMs.

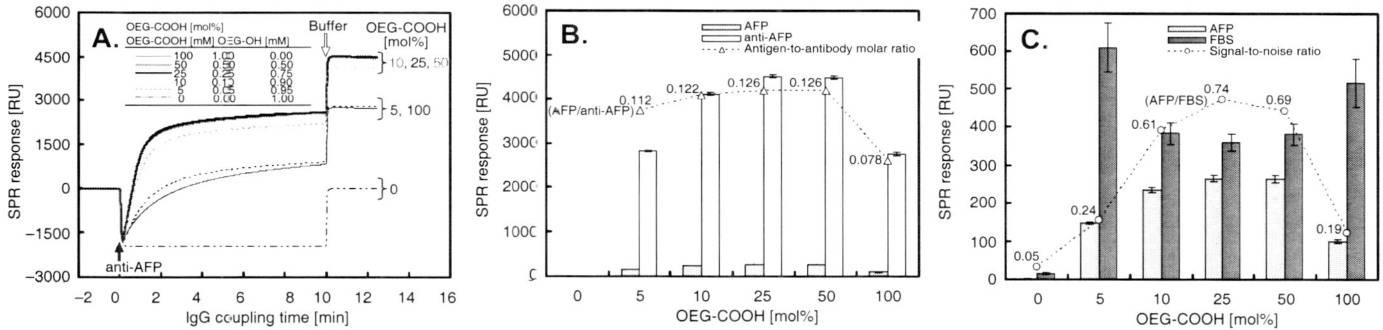

Figure 3.3 Optimization of the mixing ratios of OEG-COOH and OEG-OH. (A) Anti-AFP was allowed to react with the activated NHS ester on the OEG-mixed SAMs with various percentages (expressed as mol%) of OEG-COOH for 10 min. (B) The amount of anti-AFP immobilization and the degree of binding of 1 μg/mL AFP on the antibody-immobilized surface were used for calculation of antigen-to-antibody molar ratio (Δ) using the binding amount of AFP, anti-AFP, and their molecular weights (70 kDa for AFP, 150 kDa for anti-AFP). The degree of the nonspecific absorption of undiluted (35.6 mg/mL) FBS on the antibody-immobilized surface was used as indicators to evaluate the performance of the mixed SAM. (C) The signal-to-noise ratio (O) shows the AFP binding relative to the nonspecific adsorption of undiluted FBS. Error bars indicate standard deviations ($n = 4$).

3.2.3 Coupling Time between anti-AFP and Activated NHS Ester

The duration of the coupling reaction between anti-AFP and the activated NHS ester of the mixed SAM was examined in search of the optimal condition for the immobilization of anti-AFP. Anti-AFP was allowed to react with the activated NHS ester on the 25% OEG-COOH-mixed SAM for different times (1–20 min), and the SPR responses for anti-AFP immobilization; subsequently, AFP binding to the immobilized anti-AFP, were measured. Figure 3.4B shows that the SPR response due to anti-AFP binding reaches a plateau at 10 min, thereby indicating that the coupling time of 10 min would be enough for anti-AFP immobilization. However, the antigen-to-antibody molar ratio slightly decreases with coupling time after 4 min (Fig. 3.4B). This would raise the possibility that the number of amino groups of anti-AFP conjugated to the SAM increases with reaction time, and the excessively increased conjugation sites lower the flexibility of the immobilized anti-AFP and thereby the antigen-binding activity.

Furthermore, the AFP binding relative to the nonspecific adsorption of FBS (i.e., the signal-to-noise ratio) was found to be the highest at a coupling time of 2 min (Fig. 3.4C). This may suggest that for coupling times of <2 min, the coverage of the SAM with anti-AFP is not sufficient. For coupling times of >2 min, the redundant conjugation sites of anti-AFP due to excessively long coupling time give rise to perturbation of anti-AFP conformation, and it is possible that the subsequent exposure of the hydrophobic regions (or specific binding sites for components of FBS) of anti-AFP toward the solvent eventually leads to raised adsorption of FBS on the SAM (Fig. 3.5).

3.2.4 Comparative Analysis with Antibody-Immobilized CM5 Surface

The optimized 25% OEG-COOH-mixed SAM in this research was compared with the commercialized CM5 surface. The activation, coupling, and deactivation steps for the surface preparation were each carried out for 7 min (21 min total) or 10 min (30 min total) (Table 3.2). Under the preparation conditions of 10 min, anti-AFP was immobilized on the CM5 surface with the signal increase of

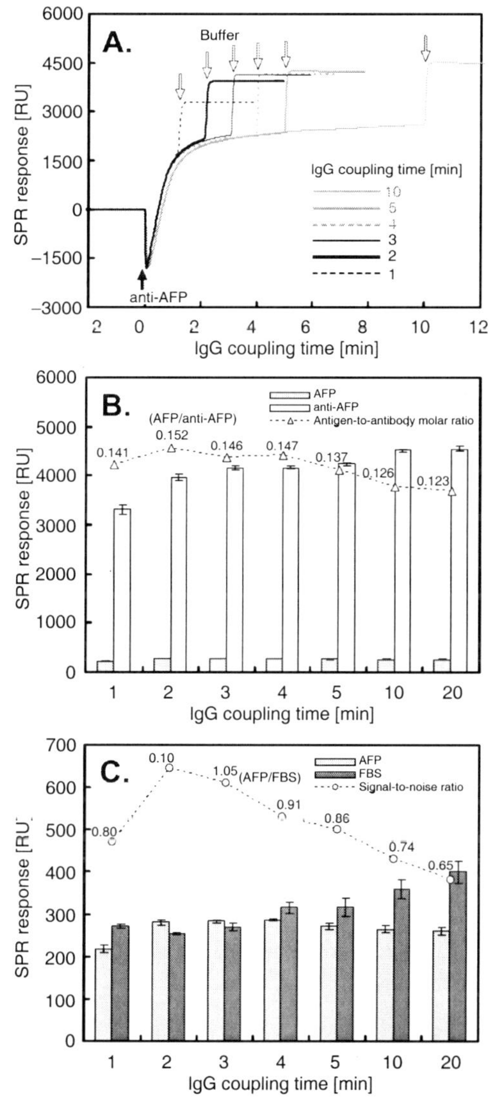

Figure 3.4 Optimization of the coupling time between anti-AFP and the activated NHS esters on a 25% OEG-COOH-mixed SAM. (A) Anti-AFP was allowed to react with the activated NHS ester on the 25% OEG-COOH-mixed SAM for different durations (1, 2, 3, 4, 5, 10 min). After the coupling reaction, an aqueous solution containing 1 M ethanolamine at pH 8.5 was rapidly injected for 10 min to block the remaining NHS ester groups on the OEG-mixed SAM. (B) and (C) refer to Fig. 3.3.

12520 RU, 3.2-fold higher in comparison with 3964 RU for the optimized 25% OEG-COOH-mixed SAM. However, the antigen-to-antibody molar ratio on the mixed SAM in the present study was 2.4-fold higher than that of the CM5 surface. The result suggests that the antigen–antibody reaction was prevented by steric hindrance between the AFP molecules and the anti-AFP molecules on the CM5 surface, probably attributable to the random orientation of the immobilized antibodies.

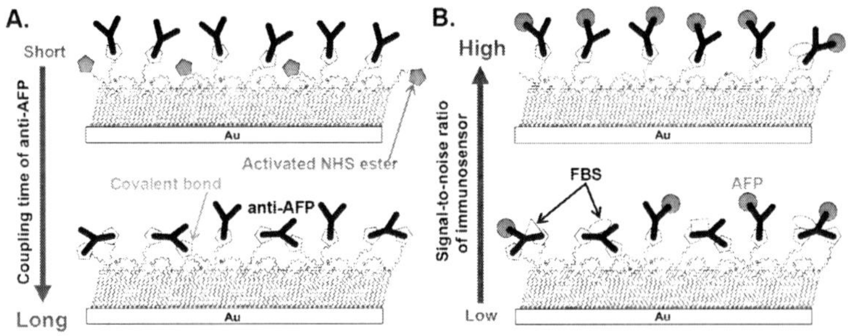

Figure 3.5 Predictive model of the coupling reaction of anti-AFP. Illustration (A) suggests that redundant covalent bonds were formed by the excessively long coupling time between anti-AFP and the activated NHS esters on a 25% OEG-COOH-mixed SAM, which induced perturbation of anti-AFP conformation on the SAM. Illustration (B) shows that reactions of (A) consequently decline in performance (expressed as signal-to-noise ratio) for the SPR immunosensor.

Moreover, the same CM5 surface exhibited nonspecific adsorption by undiluted FBS to the value of 664 RU, which was 2.6-fold higher in comparison with the 254 RU value obtained for the optimized 25% OEG-COOH-mixed SAM. However, the nonspecific adsorption was 0.63-fold lower on the CM5 surface prepared at the condition of 7 min, which equated to 7824 RU immobilized antibodies. The ratio of 0.63 (calculated by 415/664) corresponding to the decrease in the amount of nonspecific adsorption is comparable to the ratio of 0.62 (calculated by 7824/12520) corresponding to the decrease in the amount of anti-AFP binding. These results suggest that undiluted FBS adsorbs onto the anti-AFP molecules rather than onto the CM5 surface. Therefore, smaller amounts of antibody are required to capture

Table 3.2 Optimized specification of the oligo(ethylene glycol)-COOH self-assembled monolayer compared with carboxymethyl dextran[a]

Surface matrix	Surface plasmon resonace (SPR) response [RU]			Antigen-to-antibody molar ratio[b] [(ii)/(i)]	Signal-to-noise ratio [(ii)/(iii)]
	(i) Immobilization of anti-AFP	(ii) Specific-binding of AFP	(iii) Nonspecific-adsorption of FBS		
100% OEG-COOH SAM					
(10/10/10)[c]	2745 ± 35	100 ± 5.8	515 ± 64	0.78	0.19
25% OEG-COOH and 75% OEG-OH mixed SAM					
(10/10/10)[c]	4516 ± 35	265 ± 9	358 ± 23	0.126	0.74
(10/2/10)[c]	3964 ± 66	281 ± 6	254 ± 3	0.152	1.10
Carboxymethyl dextran (Sensor Chip CM5)					
(7/7/7)[c,d]	7824 ± 293	269 ± 12	415 ± 16	0.074	0.65
(10/10/10)[c]	12520 ± 304	362 ± 6	664 ± 19	0.062	0.55

[a]The representative data in Figs. 3.3 and 3.4 were summarized. To evaluate the performance of the matrix, 1 μg/mL AFP and undiluted (35.6 mg/mL) FBS were applied.
[b]The values calculated by the SPR responses (binding amount) of AFP and anti-AFP, and their molecular weights (70 kDa for AFP, 150 kDa for anti-AFP).
[c]The values in parentheses indicate the time (min) of activation/coupling/deactivation of the surface matrix.
[d]Representative conditions recommended in the protocol of Biacore Sensor Surface Handbook, BR-1005-71 Edition AB, GE Healthcare, 2008.
Anti-AFP, anti-alpha-fetoprotein antibody; AFP, alpha-fetoprotein; FBS, fetal bovine serum; RU, resonance units; SAM, self-assembled monolayer.

larger amounts of antigen for obtaining a higher signal-to-noise ratio. For this purpose, the use of a two-dimensional surface such as a SAM, and the immobilization of antibodies in a close-packed standing configuration on the SAM surface must be essential. As a result, our optimized 25% OEG-COOH-mixed SAM surface showed from 1.7-fold to 2.0-fold higher signal-to-noise ratio in comparison with the CM5 surface.

3.2.5 Conclusion

To improve the sensitivity of SPR-based biosensors, we have demonstrated effective strategies for reducing nonspecific protein adsorption on the sensor surface and maximizing antibody immobilization by using a mixed SAM composed of OEG-COOH and OEG-OH terminated OEG alkanethiols. A first strategy was to optimize the mixing condition (molar ratio) of OEG-COOH and OEG-OH in the OEG-mixed SAM, and a second strategy was to optimize the coupling time between an antibody, such as anti-AFP, and activated NHS esters on the OEG-mixed SAM. In these optimizations, to carefully observe the degree of nonspecific protein absorption was essential for high sensitivity. Our optimizations are effective for improving the orientation of antibody immobilized on the surface, would be easier than conventional methods that use CM5 (Löfås et al., 1995). In fact, the controlling of antibody orientation generally involves the induction of additional immobilized proteins such as Fc-specific antibody, protein A, G (Chung et al., 2006; Jung et al., 2007; Kausaite-Minkstimiene et al., 2010), complicated processes such as site-directed biotinylation (Cho et al., 2007; Watanabe et al., 2007). Thus, the method of antibody immobilization using OEG-mixed SAMs presented in this study has been demonstrated to be useful for biosensor surface preparation of commercial immunoassays. Our strategies are anticipated to achieve high sensitivity in those biosensing techniques such as SPR, ELISA, and quartz crystal microbalance (Pei et al., 2010; Uludağ and Tothill, 2010), which can be interfered by nonspecific adsorption.

Acknowledgments

We are grateful to Dr. Tetsuichi Wazawa (Graduate School of Engineering, Tohoku University) for critical reading of the manuscript. We are also thankful to Dr. Akio Murakami (Kobe University Research Center for Island Seas) for critical reading of the manuscript and Mr. Hiroyuki Iwasaka (Core Technology Center, OMRON Corporation) for his advice in the interpretation of the experimental data. This work was supported by the CREST (Core Research for Evolutional Science and Technology) Project of the Japan Science and Technology Agency.

Chapter 4

Nanoimprint Biosensor

The resonance condition of surface plasmon resonance (SPR) is very sensitive to the refractive index near the metal surface. According to this feature, SPR can be applied to provide a no-labeling and real-time biosensor in which the reactions of biomolecules are transduced as the change of refractive index to optical signal. As introduced in Chapter 2, some practical biosensors based on the propagating SPR with a flat gold film sputtered on a glass substrate have been developed and commercialized already. These instruments, however, require the optical setup of a total internal reflection to generate SPR. Furthermore, they need the rigorous control of flow system and inner temperature as they are also sensitive to the background noise factors in sample such as floating molecules and temperature variation. It results in commercialized SPR equipments that are large and very expensive or some cheap and small SPR equipments that are not satisfactory in terms of their sensitivity and usability.

On the other hand, LSPR (localized SPR) is expected to eliminate the ambient noise factors due to its further confinement of sensing region. Though new sensing systems based on LSPR have been keenly studied, almost all of them cannot reach a practical use yet. One of the reasons to hog-tie LSPR commercialization is the reproducibility and cost of sensor device. While very high reproducible and low-cost devices are required especially for the disposable use, there still remains challenging barriers in the

Nanoimprint Biosensors: The Fusion of Nanofabrication, Nanophotonics, and Nanobiology
Takeo Nishikawa and Satoshi Fujita

ISBN 978-981-4316-67-5 (Hardcover), 978-981-4463-42-3 (eBook)
www.panstanford.com

device fabrication of metal nanostructures for LSPR generation. For instance, metal colloids immobilized on a substrate are commonly used as a sensor substrate. To realize a uniform quality in colloid diameter and shape, high process control in deoxidization of metal ions is necessary. In addition, the uniform immobilization for avoiding aggregations and density fluctuations of colloids is still a challenge in mass production. As a stable nanofabrication method, electron beam lithography can be another candidate. The patterns are, however, produced by scanning single electron beam on a wafer and it takes a long time to complete the whole pattern that results in a high cost of devices. Other methods such as nanosphere lithography also have the difficulties in the pattern reproducibility and process throughput.

On the other side, a nanoimprint technology is focused as a novel fabrication method to produce nanostructures that are much smaller than 100 nm on a high process throughput as introduced in Chapter 1. Our group has been trying to apply this fabrication technique for preparing LSPR sensor device and to realize a small sized, low-cost and high sensitive biosensor. In the conventional LSPR, the metal free electrons are necessarily encaged in nanoscale region smaller than the diffraction limit of light since the metal configurations (colloids, nanorods, and so on) are confined in the metal island structure. For preparing the substrate by a very simple process with the nanoimprint method, the metal should be a continuous layer that is different from the conventional LSPR sensor. To realize this concept, the research was started by exploring the original nanoconfiguration that can hold LSPR characteristic with a continuous metal layer. And we have found that the periodic nanogap structure in a specific dimension can generate a standing plasmon resonance inside the nanogap. Also, the area of the intensified electric field can be easily tuned up by adjusting the dimensional parameters. The characteristics are found to be very suitable especially for the biosensor. In this chapter, the design and fabrication procedure of this device will be presented.

4.1 Design of Nanoimprint Biosensor

The eigenmodal solutions of LSPR are proved only in the case that the metal nanostructure is very simple (e.g., sphere and oval sphere). To design and analyze the plasmon resonance on the arbitrary structure, photonic simulation tools play an important role. The basic design method and result of nanoimprint biosensor are introduced in this section.

4.1.1 Simulation Methods

To simulate the optical response characteristics of nanosized objects, not only the nature of the "ray" of light but also the nature of "wave" must be considered. Two simulation methods have been used for the optical design and analysis of the nanoimprint biosensor. The dynamic analyses were carried out by using FDTD (finite-difference time-domain) method. And the static analyses were conducted by RCWA (rigorous coupled-wave analysis) method. The features and overview of these simulation methods are described in this section.

FDTD method In FDTD method, space is divided into the mesh, which is called Yee mesh. And the Maxwell's equations are solved for each mesh in each time step. This method has been used primarily to design photonic devices such as photonic crystal. The advantages of this method are that an arbitrary structure can be analyzed in principle and that rigorous dynamic simulation can be conducted based on the Maxwell's equations without any approximation. On the other hand, the accuracy depends on the fineness of the mesh size and time step. This means that larger computer memory and longer simulation period are necessary to demonstrate the simulation with higher accuracy.

To overcome this problem, continuous improvements are accumulated in this method. When the simulation model is a periodic structure, the simulation space can be enormously downsized by setting the boundary condition as the periodic one. In recent simulation tools, the mesh sizes are nonuniform and are adjusted according to the structural variations. These techniques are very helpful especially in the case where the simulation space is three dimensional and the structural patterns are too small when compared with the whole object.

In the analysis of SPR, it is necessary to simulate the behavior of the free electron in metal in particular. To realize it, some approaches are addressed. As a unique example, a large number of charged particles are prepared in the Yee mesh to simulate the electrons and ions. And the dynamic equations of these charged particles and Maxwell's equations are solved simultaneously (Usui et al., 2001). Although this method is a very promising technique since the behavior of each ions and electrons can be observed by this method, the large computer resources and simulation time is, however, inevitable.

In another method, we have simulated the characteristic of the metal by adopting the Lorentz–Drude model. The characteristics of the metal are depicted in the complex dielectric coefficient using this method. When the pattern dimension is larger than the Debye length, the simulation result can be comparatively reasonable.

RCWA method In RCWA method, the distribution of the dielectric constant of the periodic object is represented by a Fourier series expansion (Jarem and Banerjee, 1998). The static result can be calculated using this method. This has been generally used for analyzing the optical grating patterns as they are inherently periodic structure (Chambers and Nordin, 1999; Jarem, 2002). Since Maxwell's equations are solved exactly, the analysis of SPR phenomena whose simulation model includes the complex dielectric constant is possible as long as the structure is periodic. It should be noticed that the analysis accuracy is dependent on the number of harmonic expansion term in Fourier series expansion. The advantage of this method is that relatively faster simulation analysis with smaller computer memory is possible as it deals only with a static and periodic model. However, to expect higher accuracy, larger number of harmonic expansion terms is necessary, which means longer simulation time and larger computer memory are demanded. The determination of the optimal simulation parameters is sometimes required before conducting simulation.

In our work, these two methods are used complementarily. RCWA method is mainly used for analyzing the static result such as the calculation of reflection spectra. And FDTD method is used for examining the dynamic phenomena in detail.

4.1.2 Design Concept

In the conventional LSPR sensor devices, gold colloid is mainly used to generate LSPR. Though they are relatively easy to be designed and to be analyzed, the fabrication process stability and process cost are difficult issues for a practical use. As described above, we have proposed a different type of LSPR biosensor based on the nanoimprint fabrication process. The basic concept of this method is that the enhanced resonant electric field is localized around the periodic metal nanopatterns prepared by the nanoimprint and subsequent sputtering processes. Though this seems to be very simple, there are some bigger challenging points to achieve it. For instance, in the conventional LSPR, the electric field is confined in the metal island structure such as colloid, triangle pole, and so on. As a result, the localization of electric field is realized when the dimension of island structure is adequately smaller than the diffraction limit of light. If the metal film is covered continuously to the sensor surface, free electrons can basically move freely inside the film, and the confinement effect of free electrons seems to be challenging (Fig. 4.1).

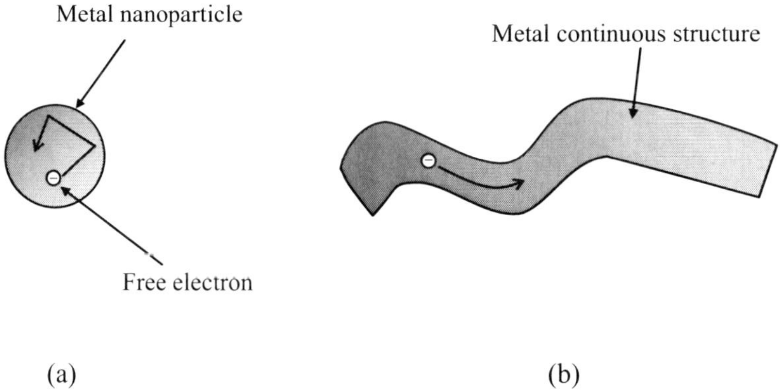

Figure 4.1 Schematics of the movement image of free electrons in (a) metal nanoparticle and (b) metal continuous structure. The free electrons in the metal nanostructure are essentially confined in it while they can freely move around in the continuous structure.

To overcome this issue, we focused on the grating type of SPR mode. In this mode, the electric field occurs between the neighboring convex patterns. We expected that the electric field will be confined between the neighboring convex patterns even if their distance is much more closed. Therefore, when the small gap structure is prepared on the metal surface, the intensified electric field is expected inside the gap. To validate this hypothesis, we have conducted a computer simulation for the typical pattern structures (Fig. 4.2).

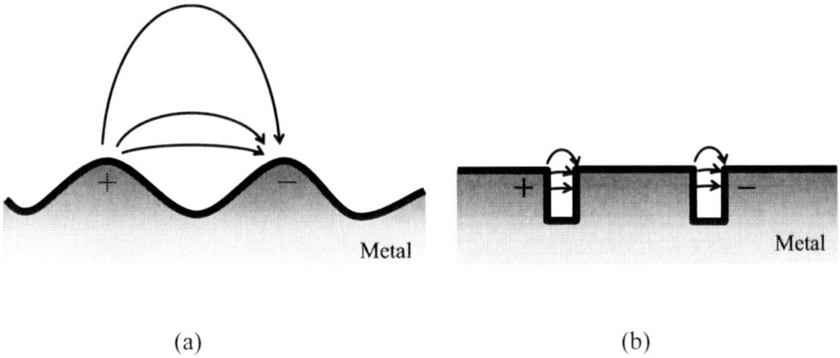

Figure 4.2 Basic concept of nanoscale confinement of the electric field on the continuous metal structure. By closing the opposed convex parts from (a) sub-micron scale to (b) nanoscale, the electric field between them will get to be confined in nanoscale.

4.1.2.1 Simulation model for basic design study

The basic structure of nanopatterns can be divided mainly into two models: convex pattern or concave pattern. The direction of incident can also be divided mainly into two modes: from the surface of the substrate or from the backside of the substrate. To observe the rough optical characteristic of these four combination models, the absorbance spectra were calculated by using RCWA method (RSoft Inc., DiffractMOD). The simulation models are shown in Fig. 4.3. The period, width, and depth (height) of the present nanopattern are 180 nm, 60 nm, and 45 nm, respectively. And gold is used for generating SPR. Its thickness was 40 nm. In RCWA method, the calculation is conducted as the periodic patterns are arranged in a row, though one pattern is shown in the model. Two-dimensional simulation is carried out to simulate the groove structure, which is uniform along the perspective direction for saving the simulation time and computational memory. The polarization of incident light is perpendicular to the groove direction in all models. The typical simulation parameters are shown in Table 4.1.

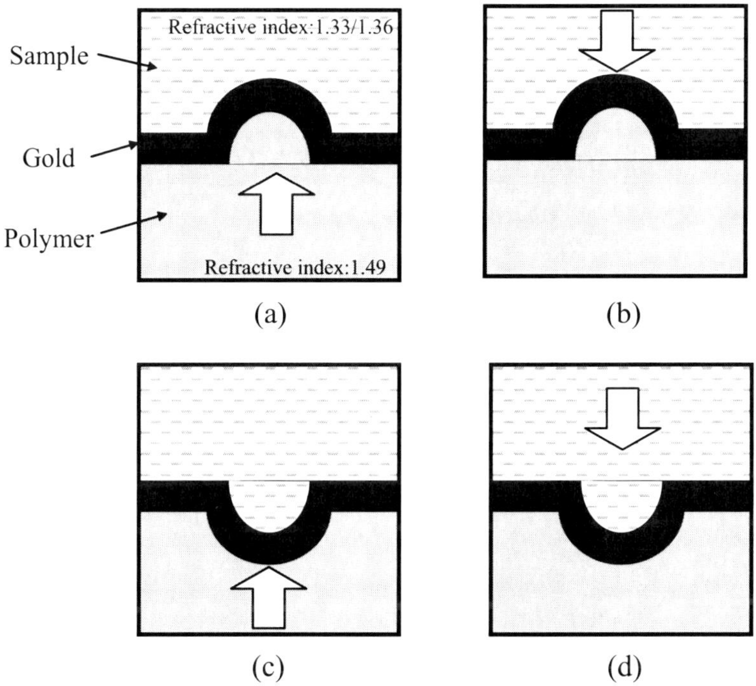

Figure 4.3 Typical simulation model schematics. (a) The nanopattern is convex and the incident light is from bottom side, (b) the nanopattern is convex and the incident light is from top side, (c) the nanopattern is concave and the incident light is from the bottom side, and (d) the nanopattern is concave and the incident light is from top side.

Table 4.1 Simulation conditions and parameters for RCWA simulation

Simulation conditions	Simulation software	Diffract MOD (R-Soft, Inc.)
	Number of PC	1
Parameters	Dimensions	2-D
	Harmonics number	30
	Model for gold	Lorentz–Drude model

4.1.2.2 Basic design of nanostructure and the direction of incident light

To estimate the sensitivity of each model, the refractive index of the sample material was changed from 1.33 to 1.36. Figure 4.4 shows the simulation result of reflection spectra in which the above four combination models are included. And the static electric field distribution at each reflection peak is also calculated by the RCWA software. This result is shown in Fig. 4.5. A detailed analysis of these results is depicted in this section.

(a) Nanopattern is convex and light enters from the backside of the substrate

The reflection peak in Fig. 4.4 is observed around 620 nm. Even if the refractive index of the sample material changes, the peak wavelength is not almost affected. In Fig. 4.5, we can see that strongly enhanced electric field, which is perpendicular to the groove structure (E_x), is inside the backside of the convex structure. These results mean that the sensor seems to be sensitive to the refractive index of the substrate material and not to the sample material. In sum, this model is not suitable for the sensor.

(b) Nanopattern is convex and light enters from the front side of the substrate

Relatively broad reflection peak is observed around 520 nm. The sensitivity to the refractive index of the sample material is 133 nm/RIU (RIU, refractive index unit), which is higher than the result of (a). In Fig. 4.5, slightly enhanced electric field is observed between the top of the convex pattern.

(c) Nanopattern is concave and light enters from the backside of the substrate

In this model, relatively broad reflection peak is observed around 540 nm. When the refractive index of the sample material changes, the peak shift sensitivity was only 17 nm/RIU. The static electric field distribution in Fig. 4.5 shows that strongly enhanced electric field barely exists around the nanopattern on the substrate side in this model. Most of light does not reach the front surface and this results in the low sensitivity.

(d) Nanopattern is concave and light enters from the front side of the substrate

The reflection peak in Fig. 4.4 is observed in the vicinity of 600 nm. The peak shift sensitivity to the refractive index of the sample material is 217 nm/RIU, which is the highest in these four simulation models. The static electric field distribution at the peak wavelength in Fig. 4.5 shows that strongly enhanced electric field exists inside the concave structure.

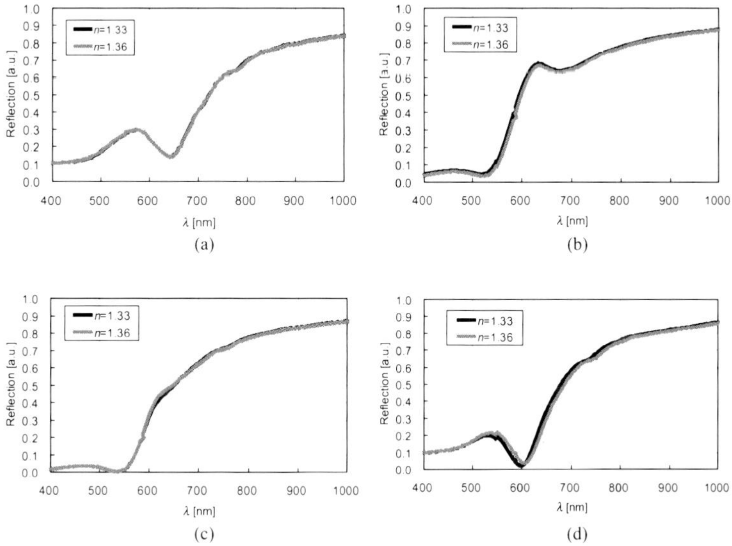

Figure 4.4 Refection spectra of the typical model calculated by RCWA simulation. (a), (b), (c), and (d) correspond to the results of models in Fig. 4.3.

According to these simulation studies, the incident light should enter from the front side of the sensor substrate as the enhanced electric field exists in the sample material in this case. Furthermore, the enhancement of the electric field is stronger in the concave structure than that in the convex structure. Considering the sensitivity of the conventional gold colloidal method is about 100 nm/RIU, the present result is promising for using a biosensor. The static electric field shown in Fig. 4.5 shows that the enhanced area is confined inside the gap of the concave structure. It means that the electric field can be localized in the area smaller than the

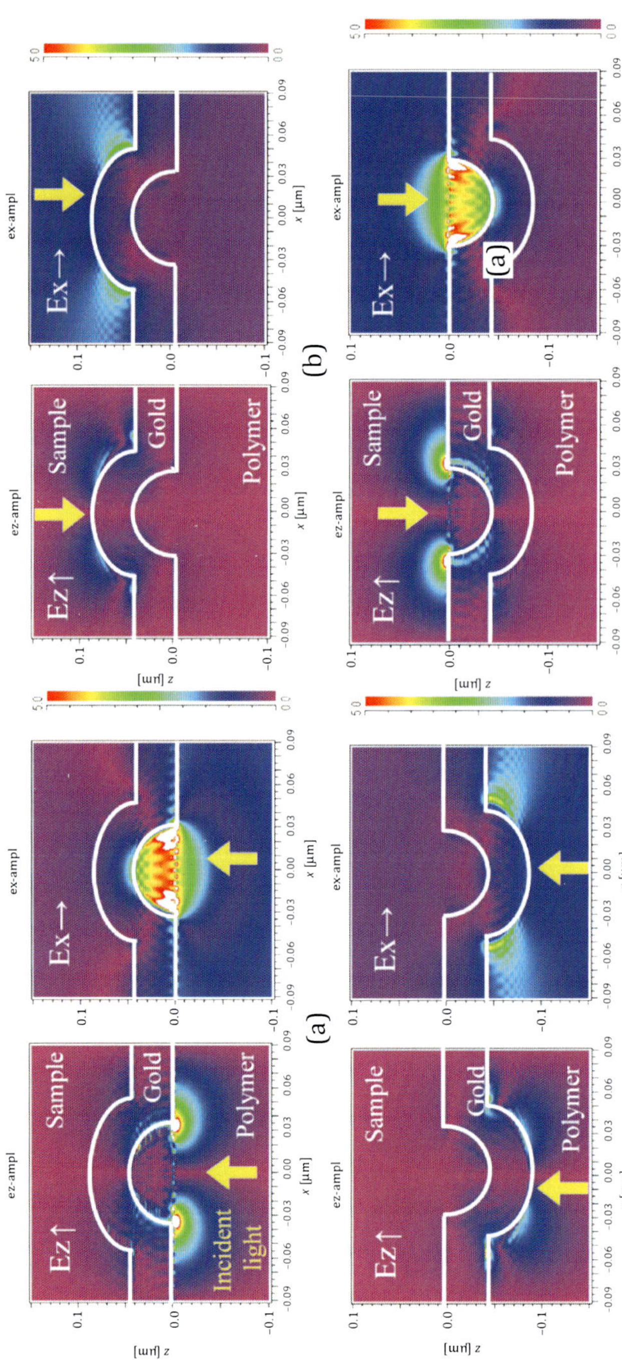

Figure 4.5 Distribution of the static electric field at the absorbance peak wavelength calculated by RCWA simulation. (a), (b), (c), and (d) correspond to the results in Fig. 4.3.

diffraction limit of light. From this result, the localization of SPR is proved to be possible by preparing nanogap structure even if the metal layer is continuously prepared on the sensor surface.

4.1.3 Design for Nanoimprint Biosensor

As described in the previous subsection, it is demonstrated that the concave, also known as "nanogap" structure, can generate the localization of SPR. In this subsection, the detailed analysis about the relation between the following structural parameters of nanogap structure and its optical characteristic is presented.

The typical structural parameters to be considered are the following five items that are given in Fig. 4.6:

- Thickness of the metal layer.
- Width of nanogap structure.
- Depth of nanogap structure .
- Angle of sidewall inside nanogap structure.
- Pattern period.

The optical characteristics to examine the biosensor are as follows:

- Resonant peak (dip) wavelength.
- Sharpness of the resonant peak.
- Peak wavelength shift to the change of refractive index of sample material.

These relationships are analyzed by using simulation as the first step.

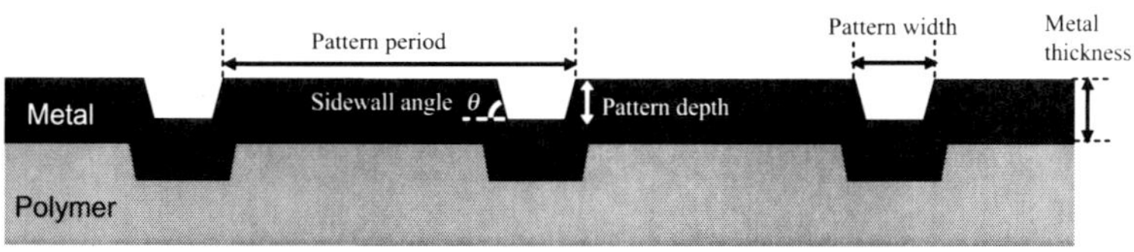

Figure 4.6 Structural design parameter for nanoimprint biosensor.

4.1.3.1 Thickness of metal layer

As a first step, the dependence of the optical characteristics on the thickness of the metal layer is studied. The simulation model is shown in Fig. 4.7. On the nanopatterned polymer layer whose refractive index is 1.49, a gold layer whose thickness is expressed as "T" is prepared. In addition, a sample layer is formed on it. In this analysis, the period, width, and depth of nanopattern on the gold layer are 280 nm, 60 nm, and 40 nm, respectively. The two-dimensional simulations are conducted on the conditions that T is 10 nm, 20 nm, 40 nm, 60 nm, 80 nm, 100 nm, 120 nm, 150 nm, and 200 nm. The incident light with a linearly perpendicular polarization to the perspective direction of nanogap structure enters from the sample material side. The reflection spectrum is calculated by using RCWA method when its wavelength is scanned from 450 nm to 1000 nm with 5 nm interval. The effects of other simulation conditions are evaluated in advance. As a result, the number of harmonic expansion terms is selected as 20 and mesh size is 0.2 nm so that the error factors should be suppressed with saturation.

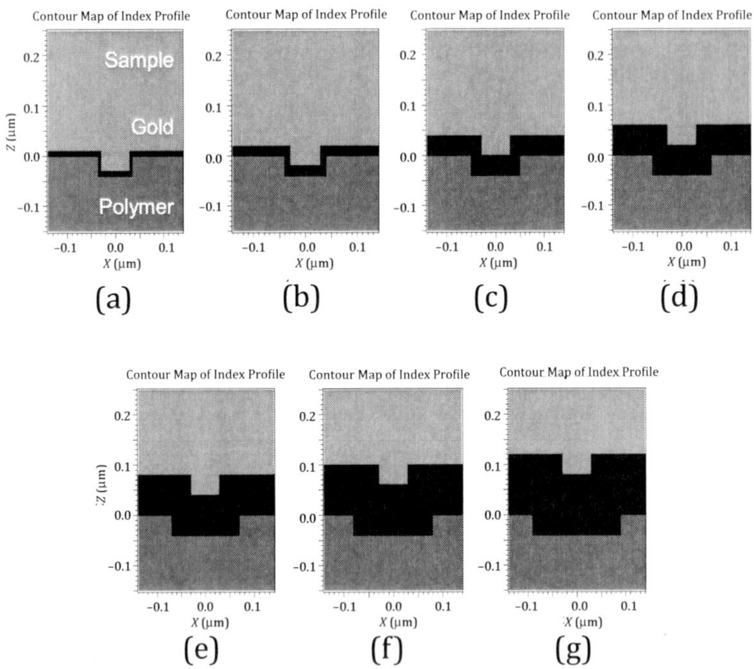

Figure 4.7 Simulation model examples for demonstrating the effect of the gold layer thickness. The thickness of the gold layer is (a) 10, (b) 20, (c) 40, (d) 60, (e) 80, (f) 100, and (g) 120 nm.

Figure 4.8 shows the reflection spectrum results. On the conditions that the thickness of the gold layer are smaller than about 40 nm, no sharp dip of the reflection spectrum was observed and the reflection coefficient was relatively low. This seems to be caused by the transmission of the incident light through the gold layer in this case. On the conditions that the thickness of the gold layer is larger than 80 nm, sharp dip of reflection coefficient caused by the generation of SPR is observed at about 660 nm. Though the optimal thickness of gold layer is about 50 nm in the conventional Kretschmann configuration SPR, this SPR model needs thicker gold layer as it does not embrace the total internal reflection system.

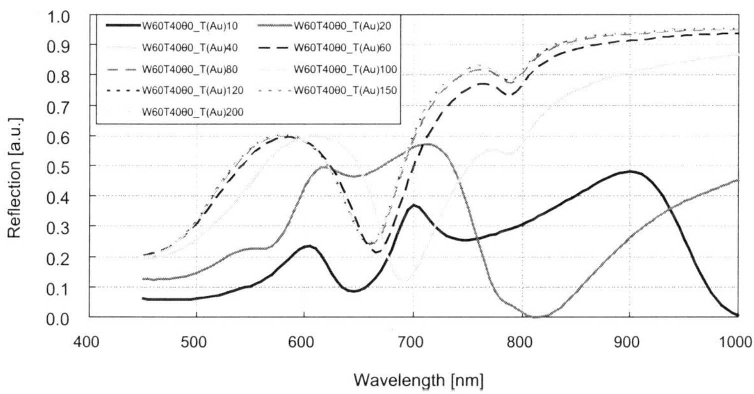

Figure 4.8 Simulation results of the reflection spectra by changing the thickness of the gold layer.

To realize a higher sensitive biosensor, the dip wavelength shift to the change of refractive index of the sample material should be larger. In the simulation model, the refractive index of the sample material is changed from 1.33 (assuming water) to 1.36 (assuming ethanol) and the dip wavelength shift is observed. To increase the resultant resolution, the detailed dip wavelength was calculated by using quadratic function approximation based on the least squares method. The normalized sensitivity to the thickness of the gold layer is shown in Fig. 4.9. From this result, it is verified that the sensitivity is low in thin gold layer region and that it is improved and saturated when the gold layer is thicker than about 80 nm. In the nanoimprint SPR model, SPR occurs on the front surface of the gold layer. Therefore, the bottom side of the gold layer does not matter when the gold layer is sufficiently thick. On the other hand, the thickness of the gold layer should not be too thick from the practical fabrication point of view since thick gold layer causes the blunt edge of nanopatterns and high process cost. In summary, following conclusion is acquired here.

The thickness of gold layer should be from 80 nm to 100 nm for a sensor device.

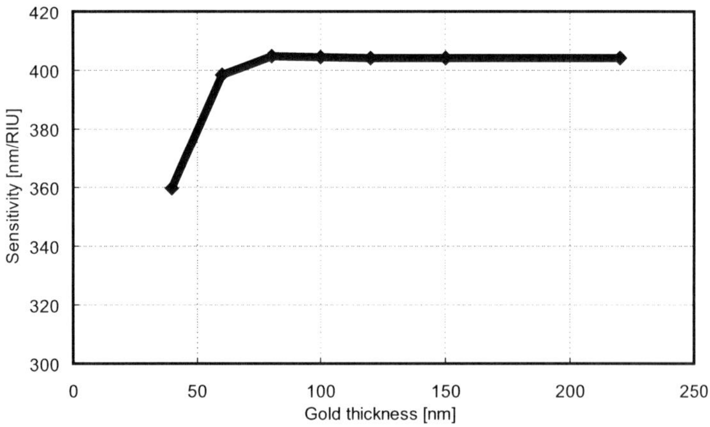

Figure 4.9 Simulation results of the sensitivity as a function of the gold thickness.

4.1.3.2 Width of nanogap structure

Optical characteristic dependence on the width of nanogap structure is verified. As a typical nanopattern, the period and depth of nanopattern are 280 nm and 40 nm, respectively. According to the previous simulation analysis, the bottom side structure of gold layer does not matter when its thickness is larger than 100 nm. To simplify the simulation, uniform gold layer as shown in Fig. 4.10 is adopted here. By this simplification, the simulation time and requisite memory in RCWA method for boundary calculation can be much reduced without increase in any error. The conditions of the incident light were the same as in the previous study. The calculated reflection spectra of sample material with the refractive index of 1.33 are shown in Fig. 4.11. And the resonant wavelength determined by the quadratic function approximation using the least squares method is shown in Table 4.2. By these results, it is verified that the resonant wavelength apparently moves to longer region when the width of nanogap structure is getting narrower. The sensitivity is also calculated by changing the refractive index of the sample material from 1.33 to 1.36 as shown in Fig. 4.12. From this result, the sensitivity is highest when the width is 20 nm and is lowest when the width is 100 nm. However, no large difference was observed when the width changed from 40 nm to 80 nm. Though it is supposed to adopt narrowest width to achieve high sensitivity, other factors such as the thickness of probe layer and target layer should be considered for using a practical biosensor. These detailed analyses will be presented later. The results are summarized as follows:

- The resonant wavelength can be adjusted by changing the nanogap width in a wide range.
- The resonant wavelength shifts to longer region when the nanogap width is smaller.
- The sensitivity is generally higher in the case where the nanogap width is smaller.

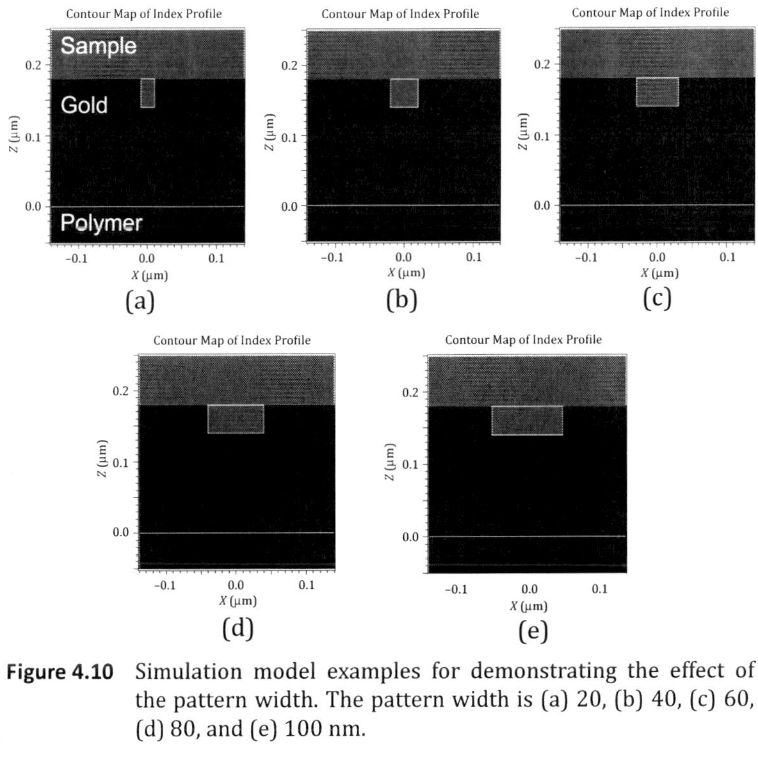

Figure 4.10 Simulation model examples for demonstrating the effect of the pattern width. The pattern width is (a) 20, (b) 40, (c) 60, (d) 80, and (e) 100 nm.

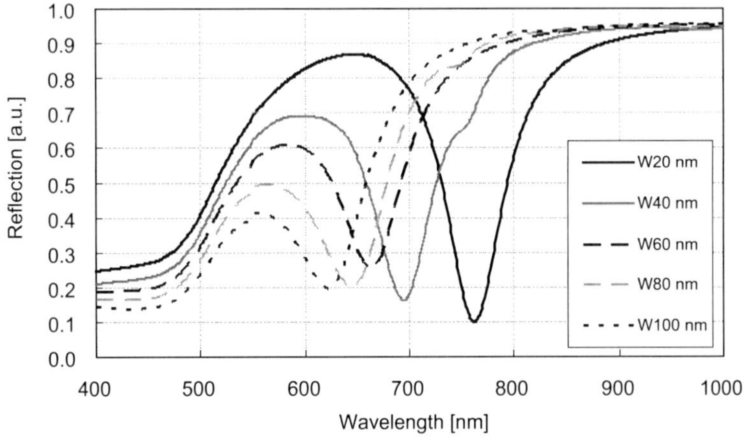

Figure 4.11 Simulation results of the reflection spectra by changing the nanopattern width.

Table 4.2 Calculated resonant wavelength and sensitivity by changing the pattern width

Pattern width [nm]		20	40	60	80	100
Resonant wavelength [nm]	@n = 1.33	761.9	694.4	663.1	643.8	621.6
	@n = 1.36	775.8	706.5	675.0	656.1	631.4
Sensitivity [nm/RIU]		464.6	402.6	397.6	410.0	325.5

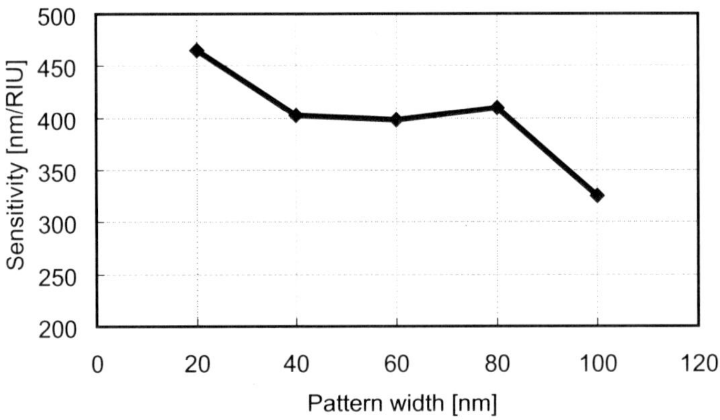

Figure 4.12 Calculated sensitivity as a function of pattern width.

In the conventional LSPR, it is known that the resonant wavelength can be changed by adjusting the material and size and structure of metal colloid. However, it is difficult to tune up the resonant wavelength in a wide range such as several hundred nanometers. In the nanoimprint SPR, it was revealed that the resonant wavelength can be easily tuned up in wide wavelength range by changing its structural parameters. And this can be one of the very important points in designing and optimizing the optical characteristics of a biosensor.

4.1.3.3 Depth of nanogap structure

The effects of depth of nanogap structure on the optical characteristic are presented here. The simulation model is depicted in Fig. 4.13. The width and period of nanogap are fixed as 40 nm and 280 nm, respectively. Other simulation parameters are the same as in the

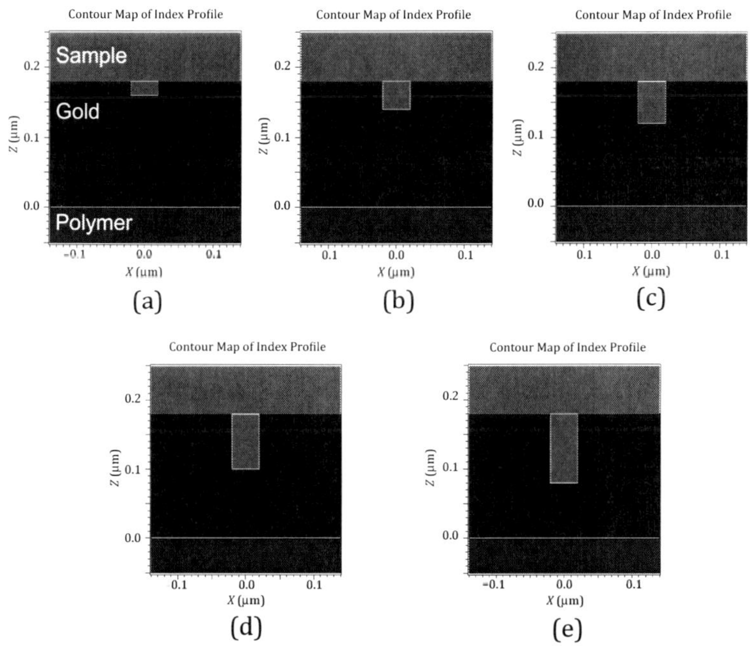

Figure 4.13 Simulation model examples for demonstrating the effect of the pattern depth. The pattern depth is (a) 20, (b) 40, (c) 60, (d) 80, and (e) 100 nm.

previous study. The reflection spectrum is shown in Fig. 4.14 and the numerical data is in Table 4.3. It is verified that the resonant wavelength apparently shifts to longer region when the nanogap structure gets deeper. The sensitivity is shown in Fig. 4.15, which indicates that the sensitivity gets higher by deepening the nanogap structure. To achieve higher sensitivity, the depth should be deeper. However, it should be notified that the resonant wavelength can overlap the absorption range of water when it gets too longer. In summary, the following results are acquired:

- The resonant wavelength gets longer when the depth of nanogap structure gets deeper.
- The deeper the nanogap structure is, the higher the sensitivity can be.

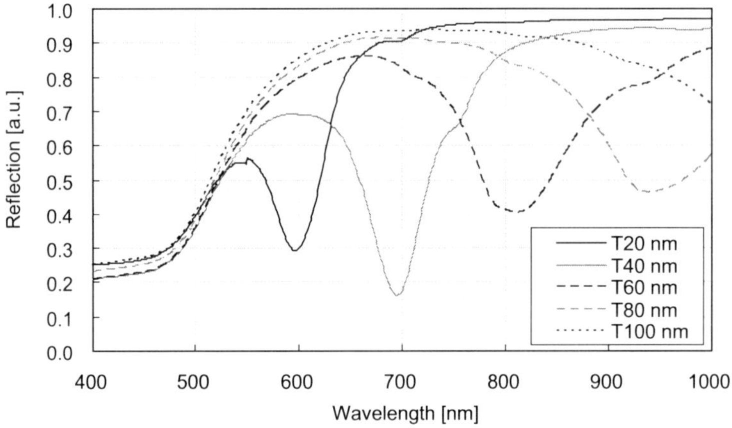

Figure 4.14 Simulation results of the reflection spectra by changing the nanopattern depth.

Table 4.3 Calculated resonant wavelength and sensitivity by changing the pattern depth

Pattern depth [nm]		20	40	60	80	100
Resonant wavelength [nm]	@n = 1.33	595.9	694.4	811.2	938.2	–
	@n = 1.36	603.8	706.5	829.0	959.8	–
Sensitivity [nm/RIU]		264.9	402.6	593.3	718.9	–

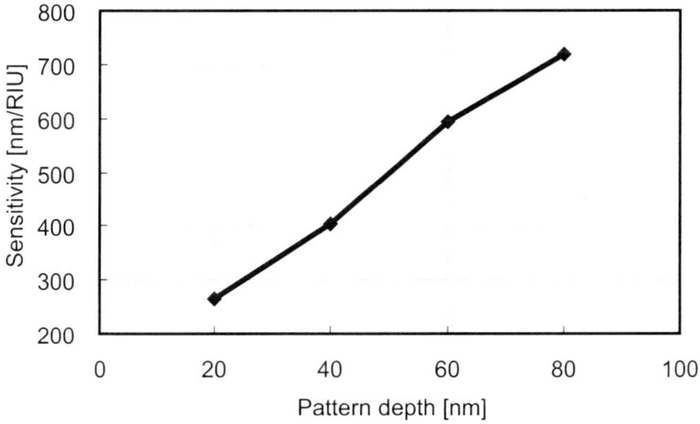

Figure 4.15 Calculated sensitivity as a function of pattern depth.

4.1.3.4 Matrix relation analysis between depth and width of nanogap structure

According to the above results, both the depth and width of nanogap structure affect the resonant wavelength and sensitivity. To optimize these two parameters, the detailed study is necessary. The analysis including the matrix relation of its depth and width is discussed here. The period of nanogap structure is fixed as 280 nm. The depth is 20 nm, 40 nm, 60 nm, 80 nm, and 100 nm and the width is 20 nm, 40 nm, 60 nm, 80 nm, and 100 nm as shown in Fig. 4.16. Other simulation parameters are the same as in the previous studies. The simulation result is shown in Fig. 4.17 when the refractive index of the sample material is 1.00, 1.33, and 1.36. The resonant wavelength and sensitivity are summarized in Tables 4.4, 4.5, and 4.6. In the previous studies, the resonant wavelength slowly gets shorter by widening the width of nanogap and it rapidly gets longer by deepening its depth. By these results, the resonant wavelength can be tuned up by adjusting the width and depth. For instance, the resonant wavelength can be in the vicinity of 620 nm in both cases of W20 T20 (when the width and depth of nanogap structure are both 20 nm) and of W100T40 (when the width and depth of nanogap structure are 100 nm and 40 nm, respectively), though their shapes are very different. The sensitivity in W100T40 is 325.5 nm/RIU, while the sensitivity in W20T20 is 276.6 nm/RIU. This result means that the sensitivity

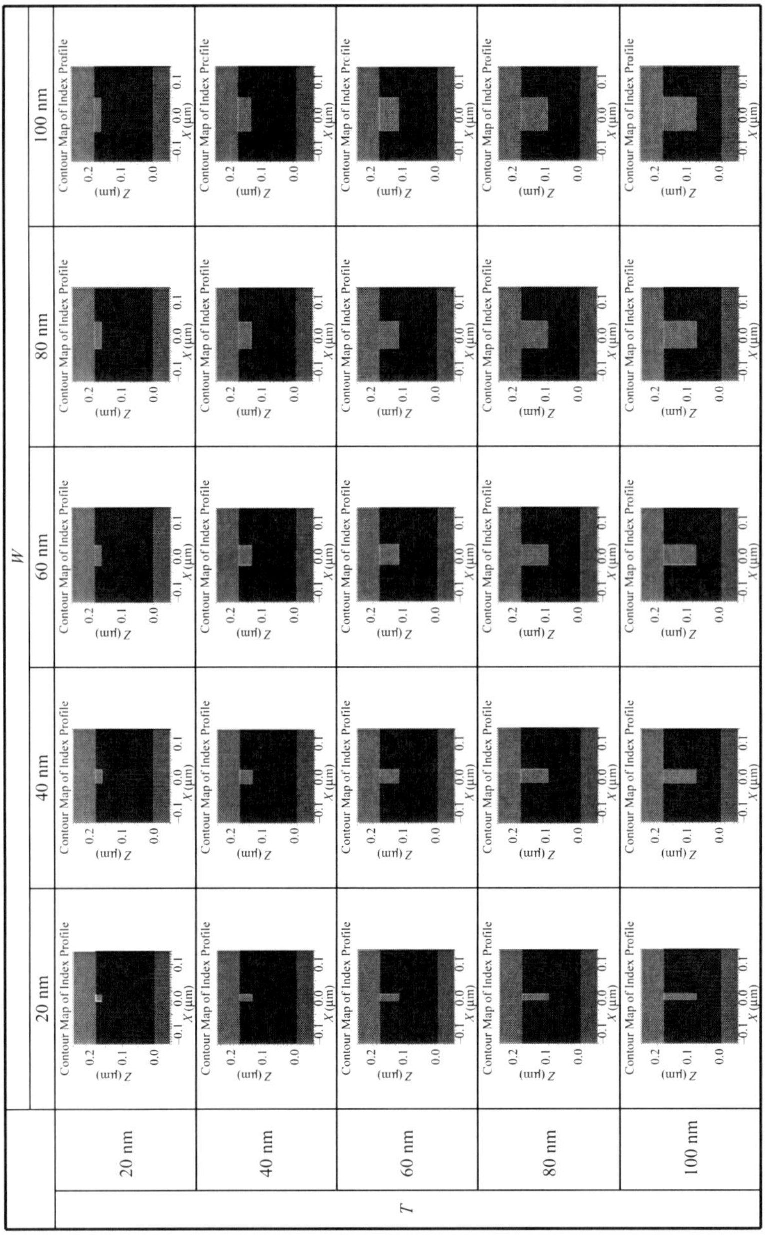

Figure 4.16 Simulation models by changing the width and depth of pattern.

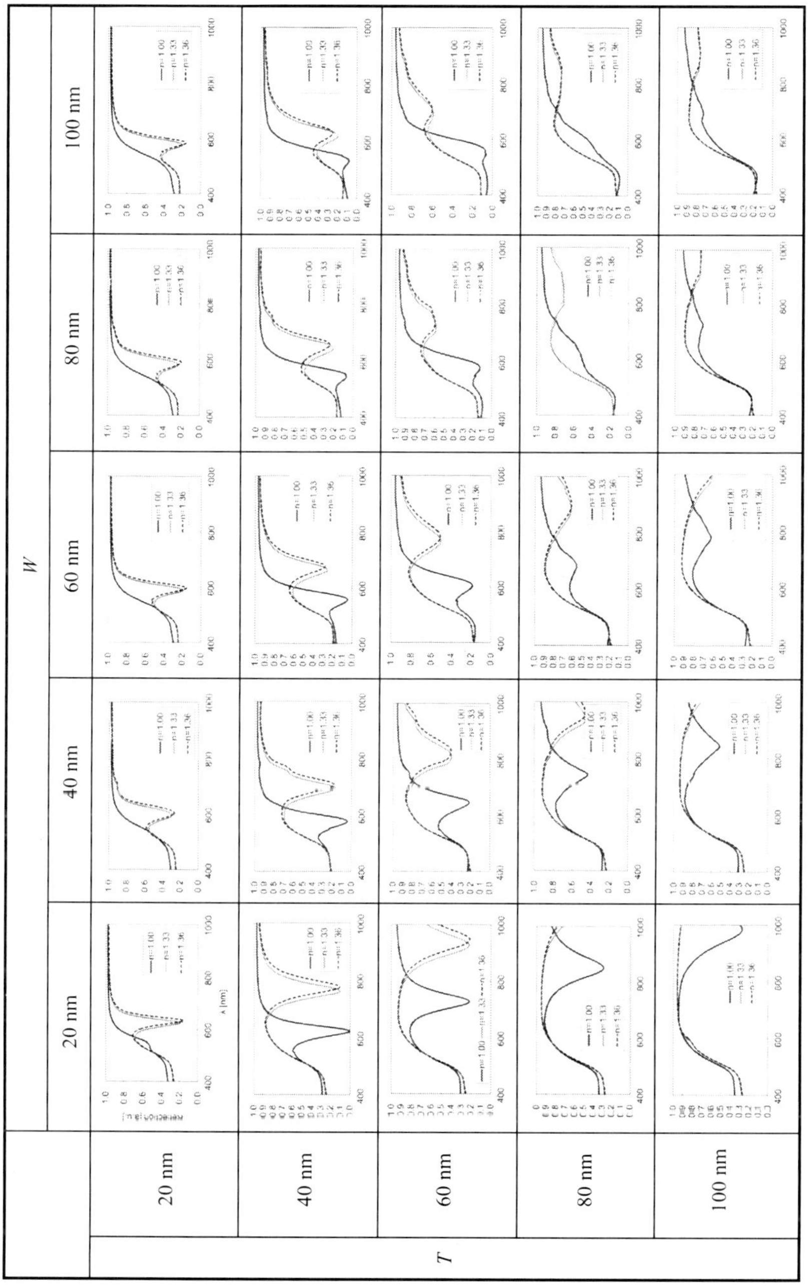

Figure 4.17 Simulation results by changing the width and depth of pattern.

of W100T40 is about 18% higher than that of W20T20. In another case, the resonant wavelengths of W60T60 and W20T40 are both in the vicinity of 760 nm. The sensitivities of W60T60 and W20T40 are 511.8 nm/RIU and 464.6 nm/RIU, respectively. Furthermore, the sensitivities of W100T60 and W40T40 whose resonant wavelengths are both about 690 nm are 446.1 nm/RIU and 402.6 nm/RIU, respectively. From these results, it can be said that the sensitivity can be increased by choosing wider and deeper structure to adjust the resonant wavelength.

Table 4.4 Resonant wavelength [nm] when the refractive index of the sample layer is 1.33

		W				
		20 nm	40 nm	60 nm	80 nm	100 nm
	20 nm	623.2	595.9	588.2	582.0	575.8
	40 nm	761.9	694.4	663.1	643.8	621.6
T	60 nm	927.9	811.2	764.8	722.8	692.0
	80 nm	–	938.2	884.4	802.1	836.0
	100 nm	–	–	–	962.2	896.7

Table 4.5 Resonant wavelength (nm) when the refractive index of the sample layer is 1.36

		W				
		20 nm	40 nm	60 nm	80 nm	100 nm
	20 nm	631.5	603.8	595.7	589.8	582.9
	40 nm	775.8	706.5	675.0	656.1	631.4
T	60 nm	946.7	829.0	780.1	741.4	705.4
	80 nm	–	959.8	902.6	856.0	853.8
	100 nm	–	–	–	984.5	913.9

The dip of the reflection spectrum curve, however, is sharper when the nanogap structure is narrower and shallower. The sharpness of the curve is also very important in considering a signal-to-noise ratio of biosensor since the noise can be larger if the curve gets broader. According to our simulation results, the curve seems to be too broad to use for a practical biosensor when the depth is deeper than 80 nm. In summary, following results are obtained.

The resonant wavelength can be adjusted by changing the width and/or depth of nanogap structure.

Even if the resonant wavelength is the same, the structure with wider and deeper gap tends to have a higher sensitivity.

The reflection curve gets broader on the above condition.

Table 4.6 Calculated sensitivity (nm/RIU) by changing the width and depth of pattern

		W				
		20 nm	40 nm	60 nm	80 nm	100 nm
	20 nm	276.6	264.9	250.6	260.0	236.2
	40 nm	464.6	402.6	397.6	410.0	325.5
T	60 nm	628.1	593.3	511.8	618.3	446.1
	80 nm	–	718.9	608.4	1797.3	590.3
	100 nm	–	–	–	742.4	575.3

4.1.3.5 Spatial period of nanogap structure

In the previous studies, the analyses are conducted on the condition that the pattern period is constant. As the next step, the effect of spatial period of nanogap structure is studied. In this model, a fixed width of 60 nm and a fixed depth of 40 nm are used. The pattern period was changed as 120 nm, 180 nm, 240 nm, 300 nm, 360 nm, 420 nm, 500 nm, 540 nm, 580 nm, 660 nm, and 780 nm as shown in Fig. 4.18. Figure 4.19 and Table 4.7 show the reflection spectrum results and the numerical data results when the refractive index of sample material is 1.33. From this result, it was found that the resonant wavelength significantly shifts to the long wavelength side by increasing the pattern period. The effect of pattern period on the sensitivity is shown in Fig. 4.20. It was verified that higher sensitivity can be acquired when the pattern period is larger. However, the resonant wavelength also gets longer to achieve higher sensitivity as in the case of pattern depth. An interesting point here is that the sharpness of the reflection curve is sharper with the longer resonant wavelength while it gets broader with the longer resonant wavelength in the case of pattern depth tuning. In summary, the following summary is acquired.

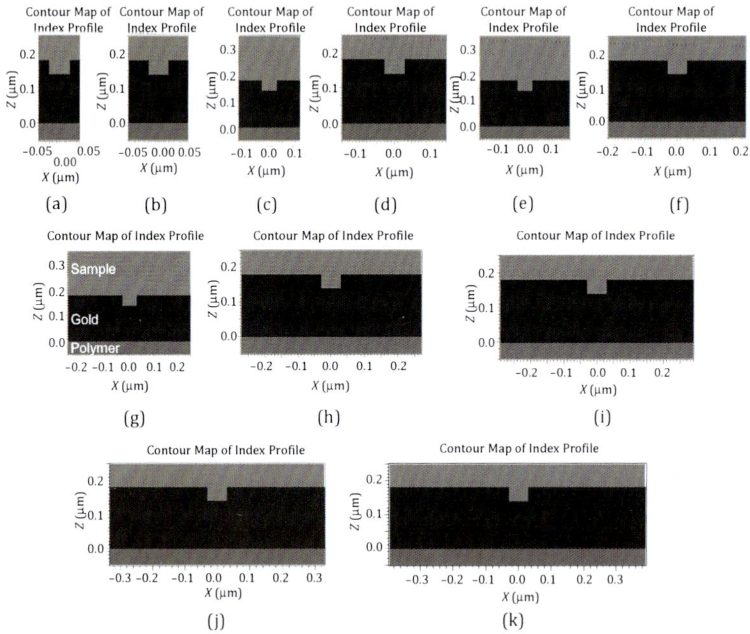

Figure 4.18 Simulation model by changing the pattern period. The pattern period is (a) 120, (b) 180, (c) 240, (d) 300, (e) 360, (f) 420, (g) 500, (h) 540, (i) 580, (j) 660, and (k) 780 nm.

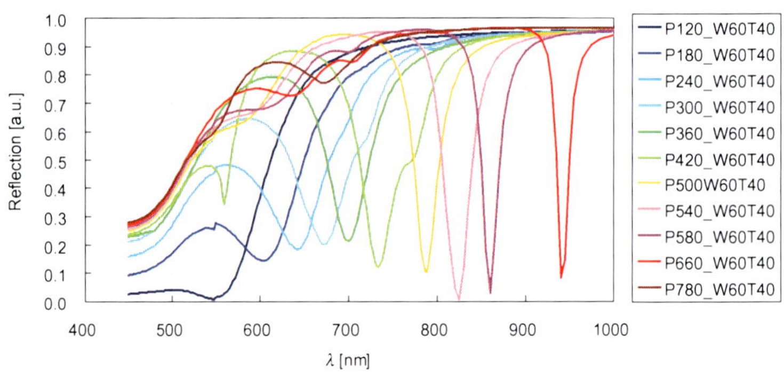

Figure 4.19 Simulation result by changing the pattern period.

The resonant wavelength and sensitivity are shifted to a longer side and raised by increasing the pattern period.

The sharpness of reflection curve is sharpened with the increase of pattern period.

Table 4.7 Resonant wavelength and sensitivity by changing the pattern period

Pattern period [nm]	120	180	240	300	360	420	500	540	580	660	780
Peak λ @ n = 1.33	545	603	642	673	700	734	787	824	861	942	-
Peak λ @ n = 1.36	546	611	653	685	713	748	804	840	880	965	-
Sensitivity [nm/RIU]	40	250	347	389	432	478	546	560	640	758	-

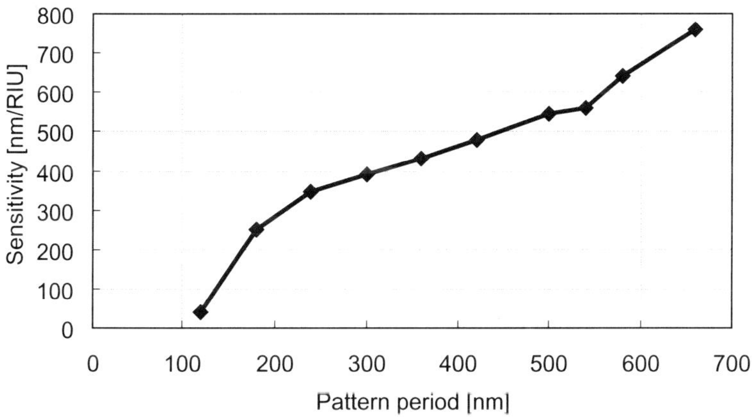

Figure 4.20 Calculated sensitivity as a function of pattern period.

4.1.3.6 The sidewall inclination of nanogap structure

So far, the simulation was carried out with the typical model in that the sidewall of the nanogap structure is right. To fabricate the nanogap structure actually, it is difficult to realize the plumb sidewall. The effect of inclination angle of sidewall on the optical properties is estimated by using simulation in this section. As shown in Fig. 4.21, the simulation model with the fixed period of 280 nm and the depth of 40 nm and width of the 60 nm was used. The width is defined at the middle position of the groove depth here. The simulation is carried out by changing the inclination angle from 60 degrees to 70, 90, and 105 degrees.

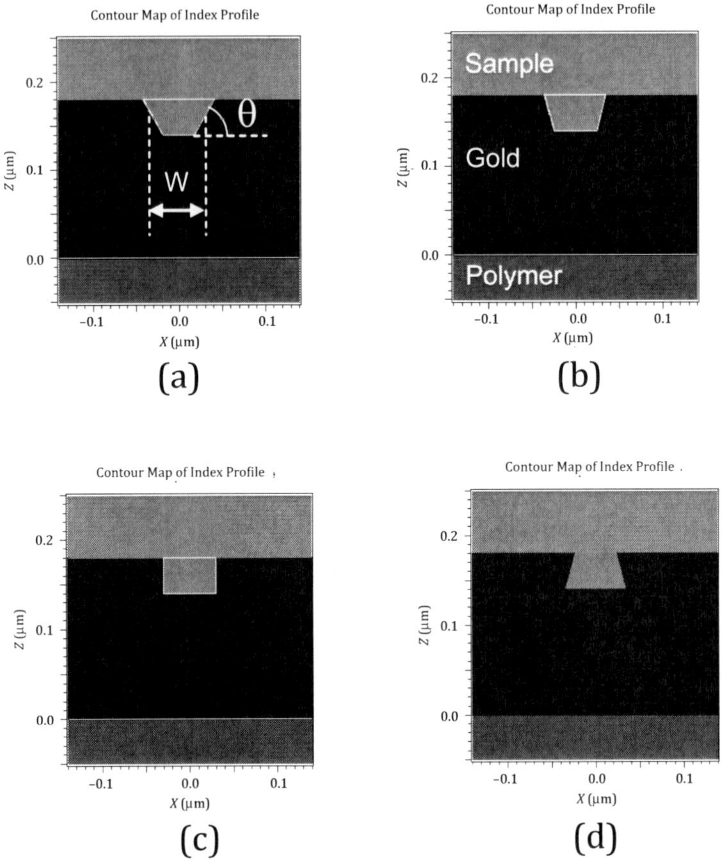

Figure 4.21 Simulation models by changing the sidewall angle.

The reflection spectrum and the numerical data results in the case where the refractive index of sample material is 1.33 are shown in Fig. 4.22 and Table 4.8, respectively. It was found that the resonant wavelength becomes shorter when the inclination angle is smaller. The sensitivity is higher when the resonant wavelength is longer as is verified in the previous studies. It is known from Fig. 4.5 that the enhanced resonant electric field is generated at the top edge of the nanogap structure. By decreasing the inclination angle, the width between edges gets larger that leads to a shorter resonant wavelength. In other words, the effect of the inclination angle can be roughly thought as the similar structural changes in the width of nanogap structure.

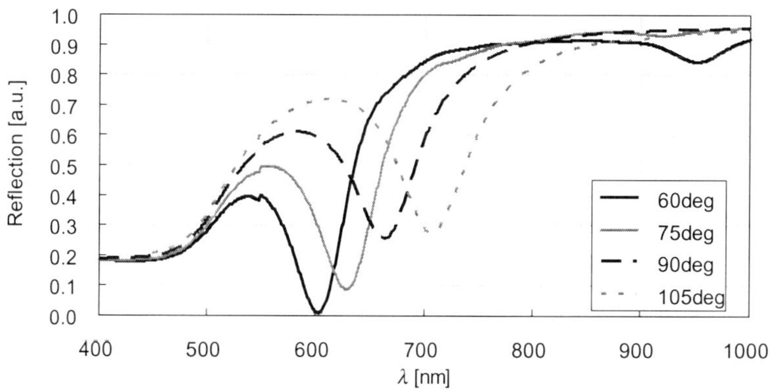

Figure 4.22 Reflection spectra by changing the sidewall angle.

Table 4.8 Resonant wavelength and sensitivity by changing the sidewall angle

Sidewall angle θ [degree]	**60**	**75**	**90**	**105**
Peak λ @ n = 1.33	602.4	628.3	663.1	706.7
Peak λ @ n = 1.36	611.1	638.3	675.0	719.3
Sensitivity [nm/RIU]	**292.5**	**332.6**	**397.6**	**420.6**

By sloping the inclination of sidewall, the resonant wavelength gets shorter and the sensitivity gets lower.

The effect of the inclination can be roughly thought as the change of width of nanogap structure.

4.1.3.7 Summary

In summary, it is verified that the width, depth, and period of nanogap pattern have significant effects on the optical properties such as resonant wavelength and sensitivity and the sharpness of the reflection spectrum curve. Their relations are shown in Fig. 4.23. From this result, it is understood that these three structural parameters correlate with each other. The sensitivity basically tends to increase when the resonant wavelength is longer. These results are the first analyses of the metal periodic nanogap structure from the point of view for applying a biosensor.

However, the evaluation of sensitivity was conducted by defining that the refractive index of the sample material changes uniformly in this study. The actual change in the biosensor occurs only in a thin region whose thickness is from a few to a few tens of nanometers corresponding to the size of biomolecules. And the changes in the other region can be noise factors. To analyze the accurate sensitivity for biosensor, the evaluation of sensitivity considering the distance from the sensor surface is necessary.

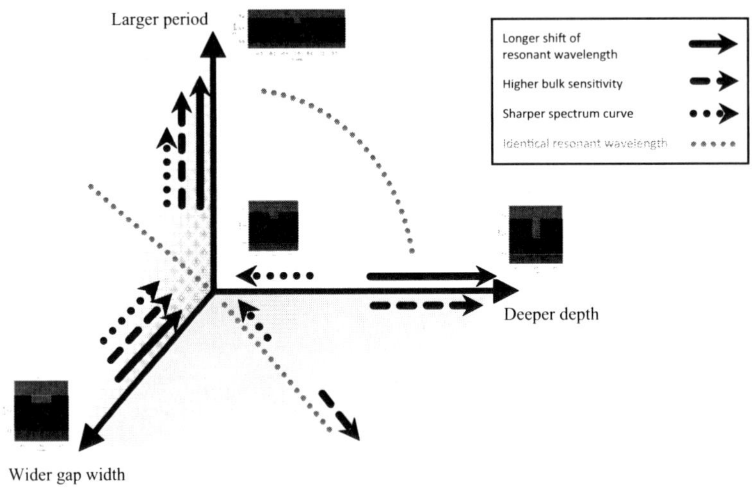

Figure 4.23 Summarized image of the bulk characteristics of the nanoimprint biosensor.

4.1.4 Optimization of Nanostructures

In the previous subsection, the general relations between the optical properties and the structural parameters are studied. To realize the practical biosensor, the more detailed analysis of the sensitivity including its dependence on the distance from the sensor surface known as "sensing depth" is necessary. The simulation method and the results are introduced in this subsection.

4.1.4.1 Simulation method to verify the sensing depth

We have used RCWA simulation software to determine the sensing depth as in the previous study. In the simulation model, nonuniform sample material is used. The refractive index of the background sample material is 1.33 and another sample material with a different refractive index which represents a "reaction layer" of biomolecules is set on the sensor surface. The thickness of the reaction layer is changed to examine the sensitivity for each thickness. In the case where the reaction layer exists on a sensor surface, the resonant wavelength shifts to the longer wavelength side due to changes in the resonance condition. However, when the thickness of the reaction layer is greater than the depth of the sensing region, the signal does not change. By examining the relation between the signal change and the thickness of the reaction layer, the sensing depth of each sensor surface can be revealed.

As the first step, the sensing depth on the conventional propagating SPR on a flat metal surface is analyzed. Figure 4.24 shows the simulation model in which 50 nm gold layer is placed on a polymer layer (n = 1.49) and a reaction layer (n = 1.36) of different thickness is put over it with the background sample material layer (n = 1.33). The incident light enters from the backside of sensor with the incident angle of 75 degrees to satisfy the total internal reflection condition. Its wavelength is scanned from 625 nm to 725 nm with the 0.5 nm increment. The simulation is carried out with the mesh size of 0.2 nm and harmonic expansion terms of 20. The reflection spectrum for each thickness of reaction layer is shown in Fig. 4.25(a). By adding the thickness of the reaction layer, it can be seen that the resonant wavelength shifts to the longer wavelength side. The shift value of the resonant wavelength with relation to thickness of the reaction layer is also shown in Fig. 4.25(b). It is shown that the shift value increases and is saturated by increasing the thickness of the reaction layer. The differential coefficient of the shift

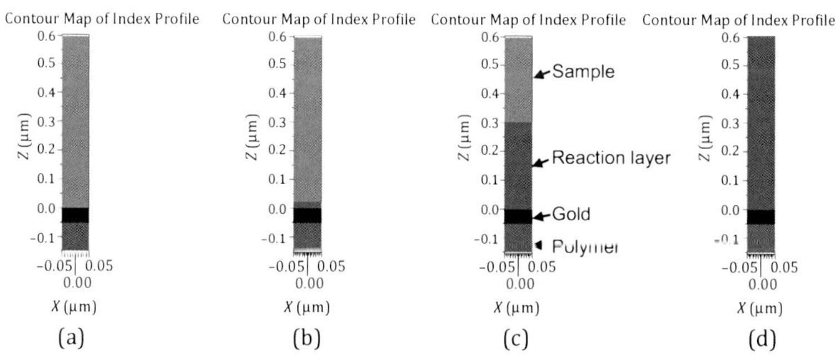

Figure 4.24 Typical simulation models for demonstrating the sensing depth of the propagating SPR. The thickness of the reaction layer is (a) 0, (b) 20, (c) 300, and (d) 800 nm, respectively.

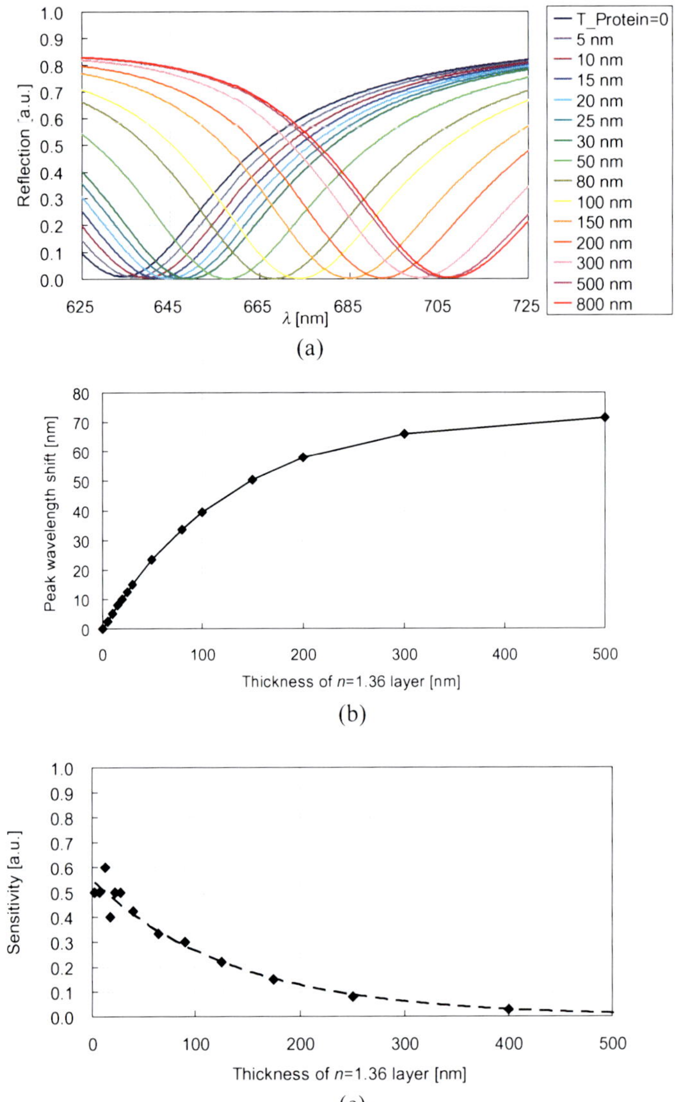

Figure 4.25 Simulation results of (a) reflection spectra, (b) resonant peak wavelength as a function of the thickness of reaction layer, and (c) the sensing depth curve.

value means the sensitivity to the change of the thickness. The calculated differential coefficient of this result is shown in Fig.

4.25(c). This value decreases exponentially by increasing the thickness. The approximated equation for the simulated curve is $y = 0.5505e^{-0.00734x}$, where y is the sensitivity (differential coefficient) and x is the distance from the sensor surface. At about $x = 273$ nm, the sensitivity becomes e^{-2} times smaller than that at $x = 0$. This result verified that the sensing depth of the conventional propagating SPR is about 300 nm, which corresponds to the theoretical value. Considering the thickness of the real reaction layer is about a few tens of nanometers, most of the sensitivity is used for the background sample material. This means that the signal tends to be affected by the background noise factors in the sample material such as the temperature fluctuation, contamination molecules, and so on. The reason why the error factor increases in the region where the reaction layer is thin in the simulation is that the change of the reaction layer (denominator) is too small to calculate the sensitivity accurately.

4.1.4.2 Sensing depth of the nanostructured surface

Consequently, the sensing depth of the nanostructured sensor surface is calculated in the same way. To use for a practical biosensor, the wavelength of the light source for sensing should be selected to be less subject to the absorption by the biomolecules in the sample bodily fluid. Furthermore, the light source device around the selected wavelength should be available easily in low cost. In this study, the structural parameters of the nanopatterns are basically chosen to satisfy that the resonant wavelength is about 650 nm for above reasons. To adjust the resonant wavelength in 650 nm when the sample material is water, a typical structure with the period of 300 nm, the top width of 60 nm, the depth of 39 nm, and the side angle of 89 degrees is used as a simulation model. The biomolecular reaction is supposed to occur uniformly along on the sensor surface, the reaction layer is stacked on the nanostructured surface as shown in Fig. 4.26. The resonant wavelength is calculated for the each model (Fig. 4.27(a)). The wavelength shift to the thickness compared with that of the conventional SPR is depicted in Fig. 4.27(b). It can be observed that the shift of the signal on the nanostructured surface is saturated at about several tens of nanometers while that on the flat surface gradually increases even at several hundred nanometers. The sensitivity on the thickness of the reaction layer compared with the flat surface is depicted in Fig. 4.27(c). The approximation curve of the nanostructured surface is $y = 0.94238e^{-0.0724x}$. The distance where the sensitivity decreases down to e^{-2} is 27.6 nm, which means the sensing depth is almost 10 times smaller than that of the conventional SPR. This result shows that the resonant electric field is confined and sensing region is localized in the significantly smaller region than the diffraction limit of light. This result also means that the sensitivity exponentially goes down by getting far from the sensor surface.

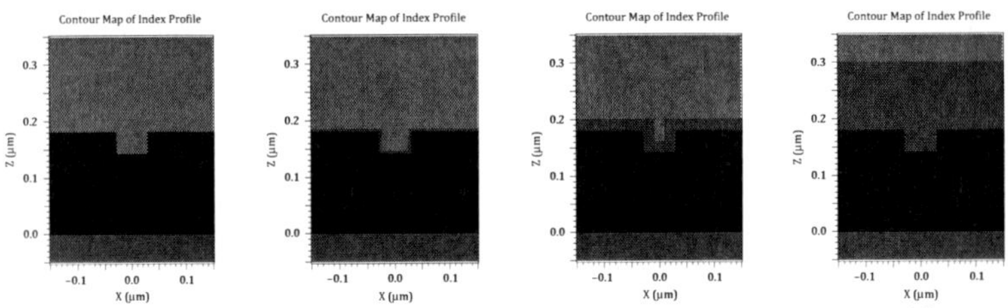

(a) (b) (c) (d)

Figure 4.26 Typical simulation models for demonstrating the sensing depth of the SPR on the nanopatterned surface. The thickness of the reaction layer is (a) 0 nm, (b) 5 nm, (c) 20 nm, and (d) 120 nm.

According to these results, the probe layer to capture the target molecule should be as thin as possible to improve the sensitivity for using a biosensor. The noise detection caused by the far region from the sensor surface is expected to be considerably eliminated while the signal might decrease in the nanostructured biosensor. This results in improving the total signal-to-noise ratio by using the nanoimprint biosensor. By applying this calculation method of the sensing depth, the relations between the sensing depth and the structural parameters are studied in the next section.

4.1.4.3 Relations between sensing depth and structural parameters

Based on the correlation of the optical properties and the structural parameters obtained in the previous subsection, the conditions whose resonant wavelength can be about 650 nm are plotted in Fig. 4.28. The angle of sidewall is assumed to be right angle here. For example, both W60T20 and W80T25 at the period of 420 nm have the resonant wavelength of about 650 nm when the refractive

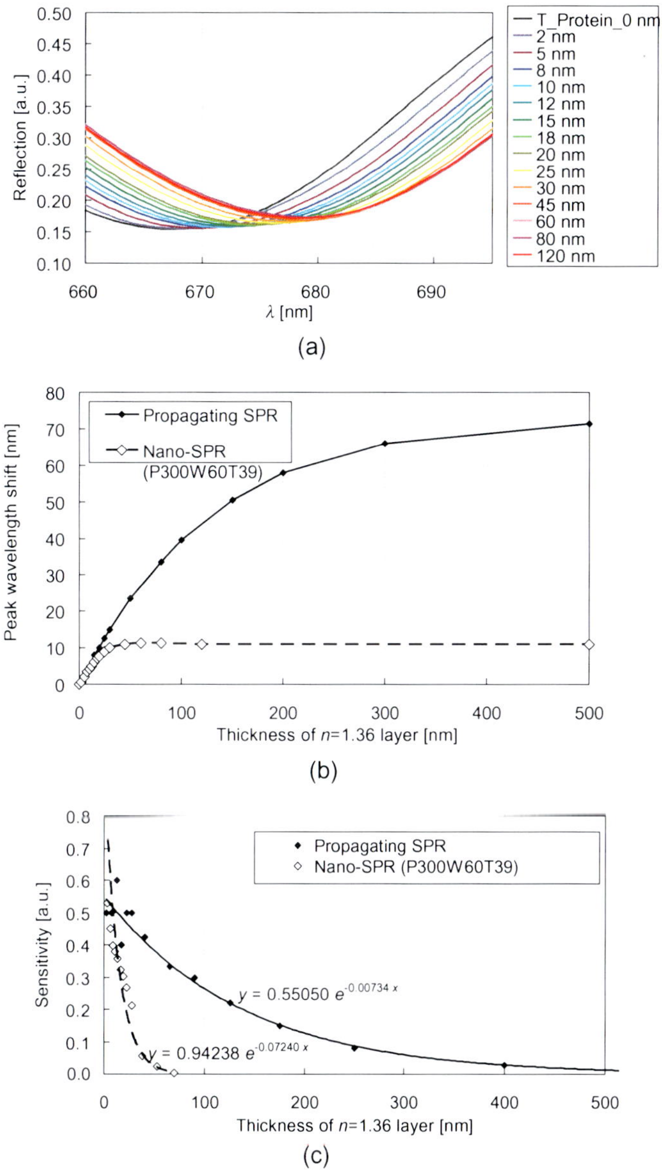

Figure 4.27 Simulation results of (a) reflection spectra, (b) resonant peak wavelength as a function of the reaction layer's thickness, and (c) the sensing depth curve of the flat surface (propagating SPR) nanopatterned surface (nano-SPR).

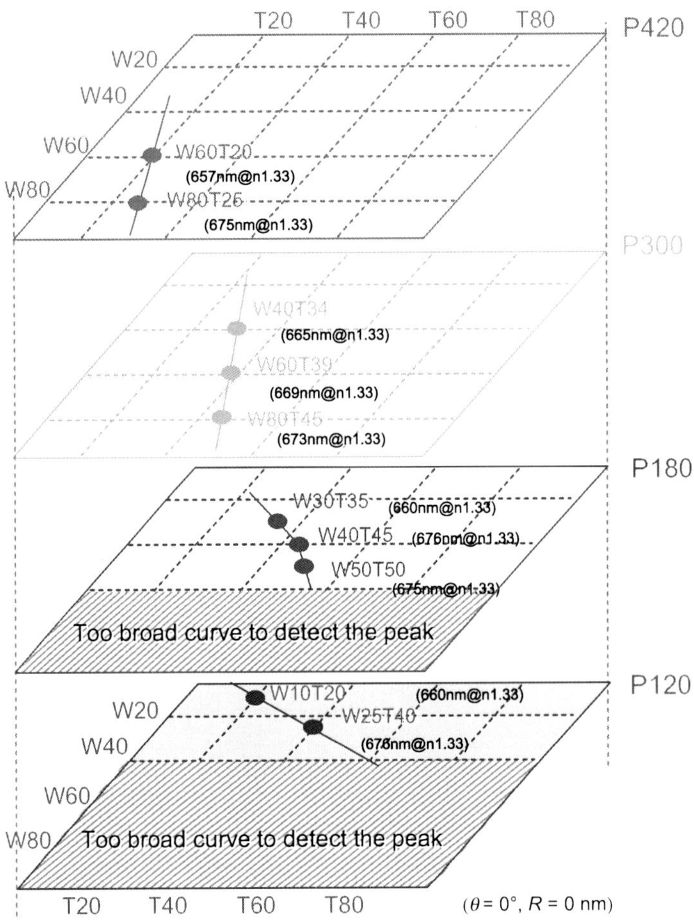

Figure 4.28 Parameter map that expresses the conditions whose resonant wavelength at n = 1.33 is about 650 nm.

index of sample material is 1.33. At the period of 300 nm, W40T34, W60T39, and W80T45 have the resonant wavelength in almost the same region. When the period is getting smaller, the curve gets too broad to detect the "dip" (local minimum) especially in the case where the width is large. From these results, it was found that the width of nanogap should be approximately smaller than one-third of the pattern period. In this section, the detailed optical properties including the sensing depth for each structure are analyzed to optimize the structural parameters.

(i) Dependence of the width and depth of nanogap structure
The effect of the width and depth of nanogap structure on the sensing depth is studied here. As verified above, the resonant wavelength can be tuned up by adjusting the width and depth of nanogap structure. The simulation models were selected as W80T45, W60T39, and W40T34 to satisfy that the resonant wavelength is about 650 nm under the condition that the period is 300 nm. The angle of sidewall is right angle. Those calculated reflection spectra are shown in Fig. 4.29(a). As described in the previous subsection, it can be observed that the calculated reflection spectrum of the narrower width has the sharper curve. The sensing depth of each model was analyzed by adding the reaction layer on the sensor surface. The result is shown in Fig. 4.29(b) in comparison with the conventional SPR. This result shows that the sensing depth slightly gets smaller when the width gets narrower. This is because the resonant electric field is generated inside the gap and the intensified electric field is more confined in it when the width is narrower. The sensitivity at the sensor surface (the distance is 0) of each model was almost regardless of the width and depth in this case.

(ii) Dependence of the period of nanogap structure
The effect of the period of nanogap structure on the sensing depth is studied. The pattern width and the angle of sidewall were fixed as 60 nm and right angle, respectively. The resonant wavelength was tuned up to be about 650 nm by adjusting the period and the depth of nanopattern. As shown in Fig. 4.30(a), the reflection spectra of P420W60T20, P360W60T32, P300W60T39, and P240W60T45 were calculated. It can be observed that the resonant wavelength is almost 650 nm in all cases and that the curve is sharper when the period is larger. The sensing depth of each model is shown in Fig. 4.30(b). This result indicates that the sensing depth gets larger with the larger pattern period. The sensitivity at the sensor surface (the distance is 0) is apparently low when the pattern period is large. As the typical examples of this analysis, the distribution of the electric field along x-direction of P240W60T45 and P420W60T20 under the resonant condition is shown in Fig. 4.31. To satisfy each resonant condition accurately, the

wavelength of incident light is selected as 665.4 nm and 657.2 nm for P240W60T45 and P420W60T20, respectively. According to this result, the electric field of P420 is more widely distributed and less enhanced than that of P240.

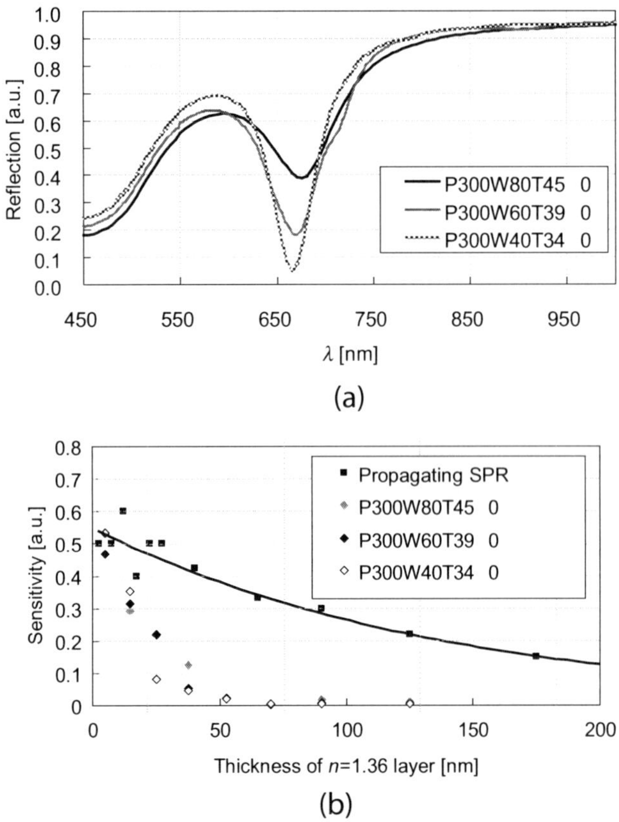

Figure 4.29 Simulation results of (a) reflection spectra and (b) sensing region when the pattern period is fixed at 300 nm.

This result means that the pattern period is one of the most important parameters to determine the sensing depth for using biosensor since it has a significant effect on the sensing depth. To use a localized SPR and to eliminate the background noise, the pattern period should be smaller since the sensitivity at the sensor surface gets higher in this case. However, we need to take account of the thickness of the probe layer to realize a practical biosensor with high signal-to-noise ratio. If the sensing depth is smaller than the thickness of the probe layer, the signal does not change with the reaction of probe layer and target molecules. To optimize the structural parameters, the pattern width and period should be appropriately selected to achieve high sensitivity while considering the thickness of the probe layer. And then the structural depth is used to tune up to satisfy the desirable resonant wavelength. The ratio of the pattern width and period should be experientially smaller than one-third since the curve gets too broad when this ratio is large. This ratio is suitable to be approximately one-fifth in our practical cases.

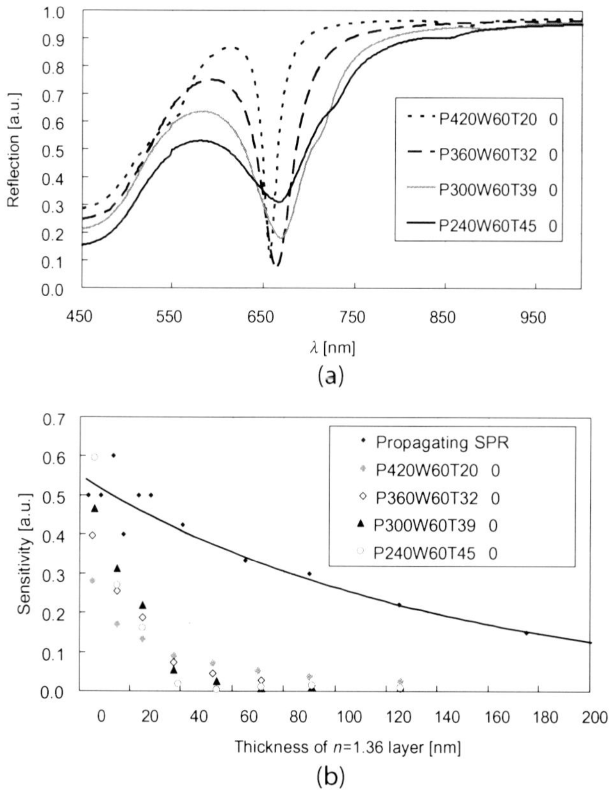

Figure 4.30 Simulation results of (a) reflection spectra and (b) sensing region when the pattern width is fixed at 60 nm.

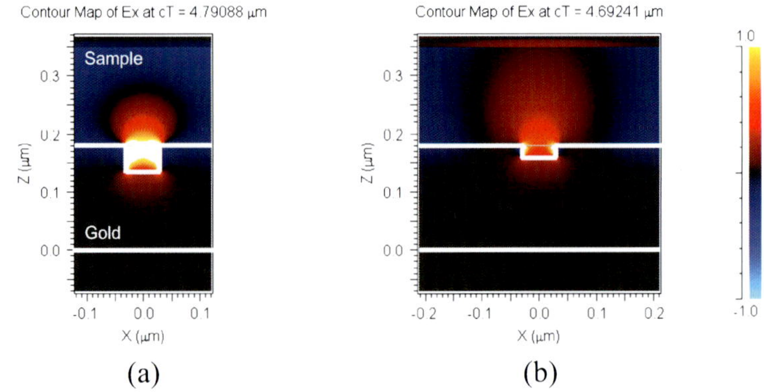

Figure 4.31 Calculated electric field distribution of (a) P240W60T45 and (b) P420W60T20 at each resonant wavelength.

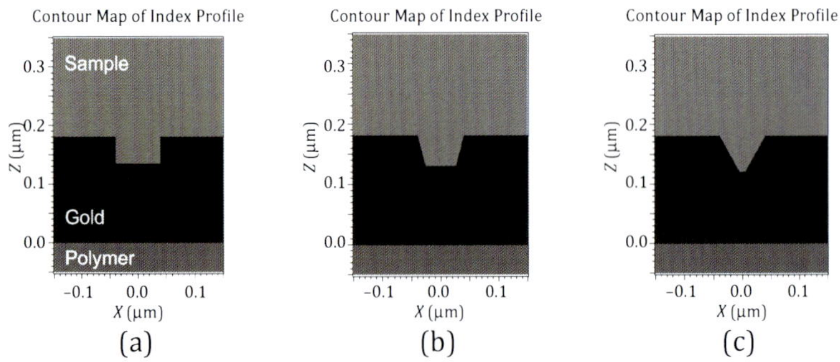

Figure 4.32 Simulation models for evaluating the effect of the sidewall angle. Sidewall angle is (a) 90 degrees, (b) 75 degrees, and (c) 60 degrees.

(iii) Dependence of the angle of sidewall

The effect of the angle of the sidewall on the sensing depth is studied here. The appropriate structural parameters whose resonant wavelength is about 650 nm under the condition that the angle is 90 degrees, 75 degrees, and 60 degrees are calculated. And the simulation models are shown in Fig. 4.32. In these simulations, the pattern period is fixed at 300 nm. The models whose top width of nanogap is 80 nm are selected. The pattern depths are adjusted at 45 nm, 50 nm, and 60 nm for 90-degree, 75-degree, and 60-degree patterns, respectively.

The reflection spectra are shown in Fig. 4.33(a). From this result, it is observed that the curve gets slightly sharper when the sidewall angle is smaller. The result of the sensing depth is shown in Fig. 4.33(b). An interesting point of this result is that the sensing depth of these different models does not have a significant difference. This means that the electric field is strongly enhanced at the top of the pattern width and does not get affected by the bottom width. The sidewall angle should have slight slope since it can provide a sharp curve.

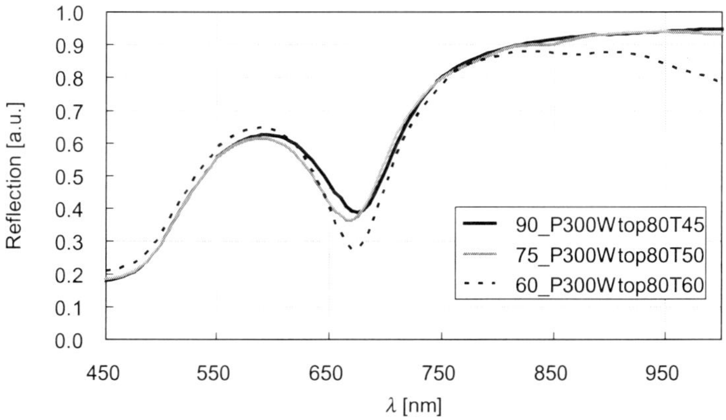

(a)

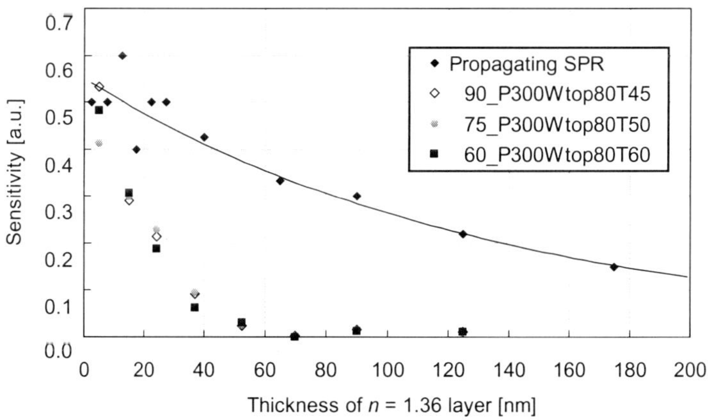

(b)

Figure 4.33 Simulation results of (a) reflection spectra and (b) sensing region by changing the sidewall angle.

(iv) Optimization examples of sensing depth for a biosensor

Based on these results, the optimization examples for a biosensor are introduced. As described above, one of the most significant factors to design the nanostructure for optimizing the sensing depth is the thickness of the probe layer. In the case where the thickness of probe layer is 20 nm (Orla18: 5 nm, antibody: 15 nm), the thickness of target molecule (antigen) is 5 nm and the effective sensing distance from the sensor surface is from 20 nm to 25 nm. As an example, the sensitivity model of P300W60T39 is shown in Fig. 4.34. It can be seen that the average sensitivity for the target molecule, known as "target sensitivity" in this case, is about 0.26 a.u. Figure 4.35 shows the target sensitivity for the simulation models with various pattern period and width when the thickness of the probe layer is 20 nm. The sidewall angle is 89 degrees and each depth is determined to make the resonant wavelength to be about 650 nm. This result verified that the target sensitivity is highest at P300W80 whose depth is 45 nm. Therefore, we could understand that P300W80T45 is a desirable structure with the probe layer of 20 nm thickness.

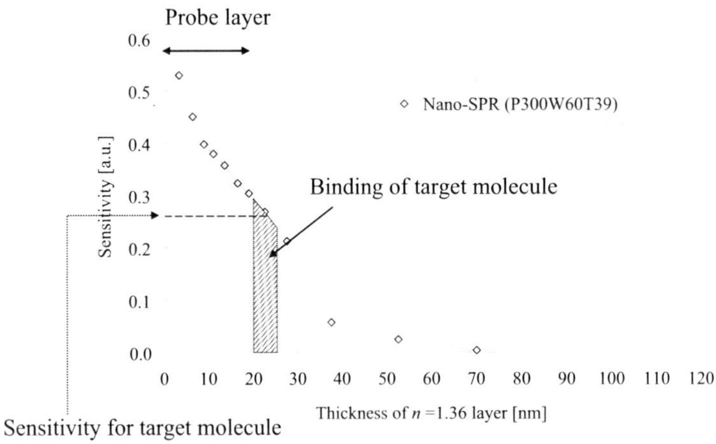

Figure 4.34 Definition of the sensitivity for the target molecule.

To improve the sensitivity, thinner probe layer is better. The probe layer can be as thin as 10 nm by using a single-chain antibody or fusion proteins. The target sensitivity in this case is also studied and the result is shown in Fig. 4.36. It is proved that the target sensitivity is wholly improved by using a thinner probe layer. And the nanostructure with the period of 240–300 nm and width of 40 nm is an optimized structure when the probe layer is 10 nm thick.

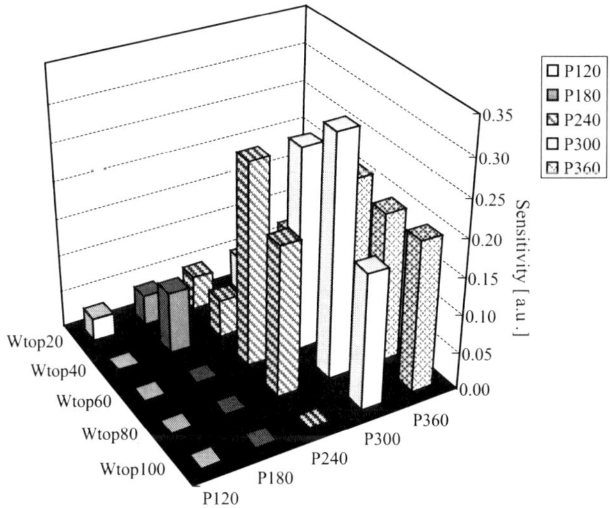

Figure 4.35 Calculated sensitivity for the target molecule on the probe layer whose thickness is 20 nm.

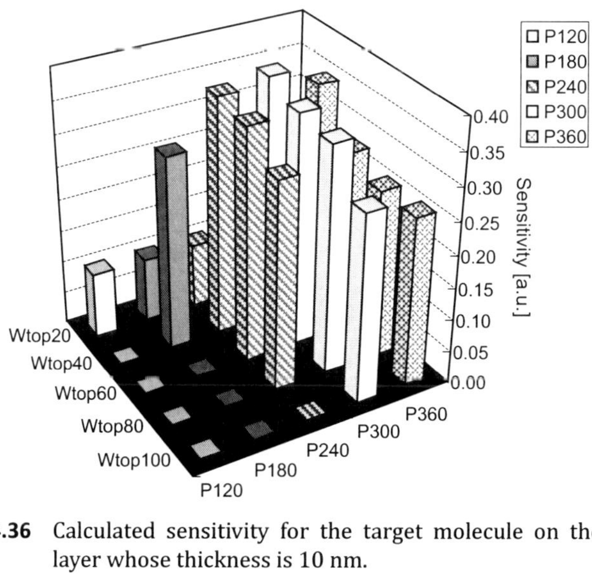

Figure 4.36 Calculated sensitivity for the target molecule on the probe layer whose thickness is 10 nm.

In this subsection, it is proved and demonstrated that the detailed analysis and optimization of nanostructure considering the sensing depth and target sensitivity can be possible by using these simulation techniques.

4.1.5 Additional Analysis of Nanoimprint Biosensor

To understand the details, additional analyses of nanoimprint biosensor based on SPR are presented in this section.

4.1.5.1 Analysis of sensitivity dependence on the position of sensor surface

In the conventional propagating SPR on a flat metal surface, the resonated electric field distribution on the sensor surface is homogeneous and is attenuated exponentially as a function of the distance from the surface. On the other hand, in the nanoimprint SPR which occurs on the nanopatterned metal surface, the resonated electric field distribution is not homogeneous and is dominated by the nanopatterns. Therefore, strictly speaking, it is expected that the sensitivity on the sensor surface is dependent on the position of the sensor surface.

To verify this sensitivity characteristic in a computer simulation, the signal change is obtained when an artificial protein is bound to each point of the sensor surface. In a simulation model, the pattern period, the top width, and depth of nanogap are 280 nm, 70 nm, and 40 nm, respectively. On its surface, an artificial protein whose diameter is 3.8 nm and whose refractive index is 1.57 is attached (Ku and Stroeve, 2004). As studied in the previous subsection, the nanostructure is directly prepared on a gold layer of 100 nm thickness for the sake of simplicity. Since the boundary condition of the simulation is periodical one, the artificial protein is supposed to be present at 280 nm intervals. As the simulation model is two dimensional, this model represents that a cylindrical protein bounds on the sensor surface with a groove structure. The artificial protein is prepared on the following six points on the nanopatterned sensor surface (Fig. 4.37):

(i) At the center position of the nanogap bottom (bottom)
(ii) At the edge of the nanogap bottom (bottom-edge)

(iii) At the center position of the sidewall (wall)
(iv) At the edge of the nanogap top (top-edge)
(v) At the top surface near the nanogap (top)
(vi) At the center position between the nanogaps (top-boundary)

The refractive index of the background sample material is uniformly 1.33. The mesh size and the harmonic expansion terms are 0.2 nm and 20, respectively. The incident light with a linearly polarization perpendicular to the nanogap structure enters from the sample material side vertically to the sensor surface. The wavelength of the incident light is scanned, and the resonant wavelength is detected before and after the artificial protein binding. And the signal shift is calculated.

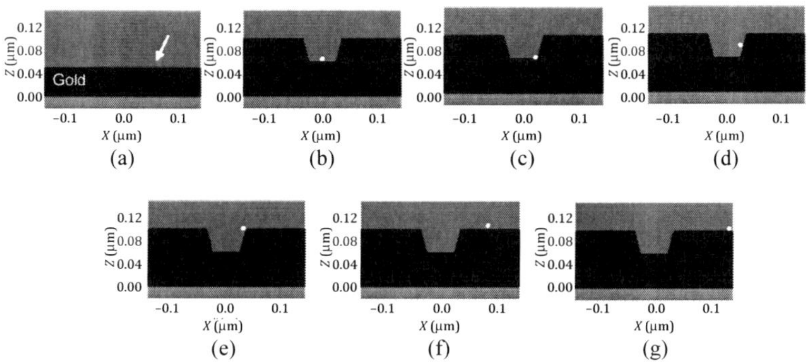

Figure 4.37 Simulation models with the dummy protein on (a) flat gold surface, at (b) bottom, (c) bottom-edge, (d) wall, (e) top-edge, (f) top, and (g) top-boundary.

To compare with the conventional SPR, a simulation model in which an artificial protein is placed on a flat metal surface is also demonstrated. In this model, a gold layer of 50 nm thickness is prepared on a polymer layer whose refractive index is 1.49. The incident angle of the light is fixed at 75 degrees and the wavelength is scanned to detect the resonant wavelength shift. Other simulation parameters are basically identical to those of the nanopatterned model.

The result of a wavelength shift caused by the protein binding is shown in Fig. 4.38. In the conventional SPR, the signal shift of 0.189 nm was observed. In the nanoimprint SPR, it was verified that the signal shift was significantly different according to the binding positions of the artificial protein. The signal shift at "top-edge" is about four times higher than that of the conventional SPR while the signal shift at "bottom" and "top-boundary" is almost 0. It is observed that "top-edge" can be hot spot of the enhanced electric field. These signal distribution corresponds to the distribution of the resonant electric field intensified by the generation of nanoimprint SPR.

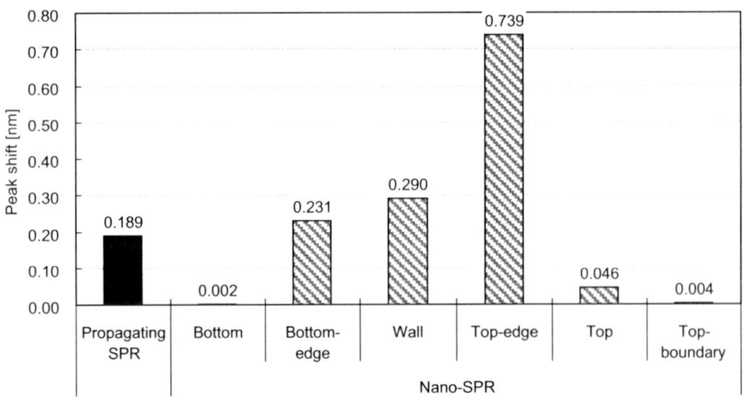

Figure 4.38 Simulation result of the signal change by dummy protein models.

Ideally, the probe layer should be prepared selectively on the hot spot area to prevent the target proteins from binding on the area where the sensitivity is low. On the present sensor surface, however, the binding of the target protein occurs uniformly since the probe layer is prepared on the all position. The signal shift is averaged among the sensitivity distribution and is assumed to be almost the same as the conventional SPR. And the noise factors in the nanoimprint SPR are supposed to be much lower than the conventional SPR. As a result, the total signal-to-noise ratio of the nanoimprint SPR can be higher than the conventional SPR.

4.1.5.2 Dispersion analysis of SPR mode

To understand the characteristic of the nanoimprint SPR, the dispersion relation is analyzed in this section.

(i) Dispersion relation of grating SPR

As a comparison, the dispersion relation of the traditional grating SPR on the metal surface is simulated. To generate the SPR by coupling the incident light to the SPR mode of metal surface, following condition needs to be satisfied by using a total internal reflection configuration.

$$k_x = k_0 n_\mathrm{p} \sin\theta = k_0 \sqrt{\frac{\varepsilon_\mathrm{m} n_\mathrm{a}^2}{\varepsilon_\mathrm{m} + n_\mathrm{a}^2}}$$

where k_x is the wave number in x-direction and k_0 is the wave number of incident light, n_p is the refractive index of the prism, θ is the incident angle, ε_m is the dielectric coefficient of the metal, and n_a is the refractive index of the sample material. Another method that does not use a total internal reflection method is the optical grating (Cullen et al., 1987; Homola et al., 1999).

In this method, the following equation needs to be satisfied:

$$n_\mathrm{a} \sin\theta + m\frac{\lambda}{\Lambda} = \pm\sqrt{\frac{\varepsilon_\mathrm{m} n_\mathrm{a}^2}{\varepsilon_\mathrm{m} + n_\mathrm{a}^2}}$$

where m is the order of diffraction light, and Λ and λ are period of nanopattern and wavelength of light, respectively. The dispersion relation is calculated by using RCWA method. The simulation model with the sine curve structure whose amplitude is 22.5 nm at an interval of 300 nm on a gold layer is used as shown in Fig. 4.39. The light enters from the material side into the gold surface with the grating. And the absorption spectra were determined with scanning the incident angle and the wavelength. The absorbance spectra of θ–λ relation are converted to that of k–ω relation by using a transfer program based on the following equation:

$$X = \frac{k}{2\pi} = n \cdot \sin\left(\frac{\theta}{\lambda}\right)$$

$$Y = \frac{\omega}{c} = \frac{2\pi}{\lambda}$$

where c is the speed of light and n is the refractive index of sample material. The result is shown in Fig. 4.40. In these equations, the relation between X and Y can be expressed by the following equation:

$$Y = \frac{2\pi \cdot X}{n \cdot \sin\theta} \geq \frac{2\pi \cdot X}{n}$$

Due to the above inequality, the region of $Y < (2\pi \cdot X)/n$ cannot be computable. The contour shows the intensity of absorbance. In this result, strong absorbance line from $(X, Y) = (1.6, 7.5)$ to $(0, 12)$ is observed. By drawing the theoretical absorbance dispersion relation as the white line in Fig. 4.40, it is observed that the theoretically calculated grating SPR mode is overlapped on the simulation result. This means that the observed strong absorption mode corresponds to the first-order grating SPR mode. This good accordance proved that the mode analysis using RCWA method is pertinent.

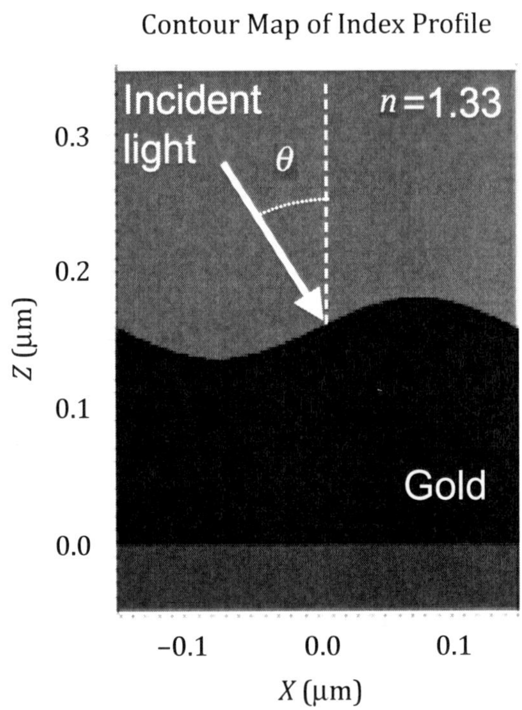

Figure 4.39 Simulation model for calculating the dispersion relation of grating SPR mode.

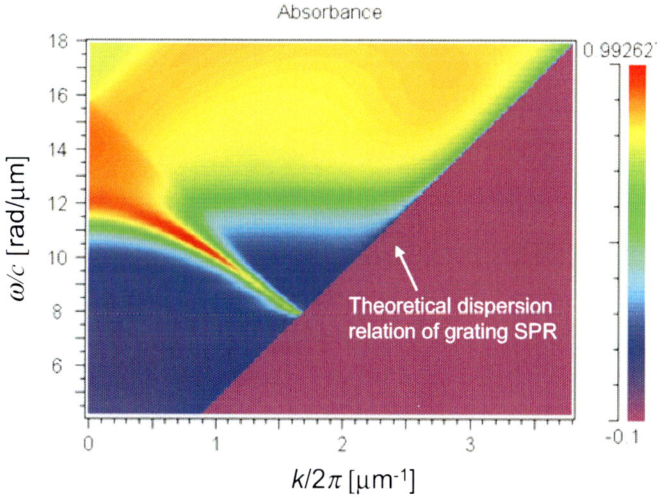

Figure 4.40 RCWA simulation result of the absorbance intensity (color contour) on a grating surface expressed as ω-k dispersion relation. The white line is the theoretical dispersion relation of the grating SPR, which is calculated based on the theoretical equations.

(ii) Dispersion relation of nanoimprint SPR

As the next step, the dispersion relation of the nanoimprint SPR is analyzed in the same way. As a simulation model, the gold surface with the nanogap structure whose width and depth are 80 nm and 45 nm is used as shown in Fig. 4.41. The pattern period and the sidewall angle are 300 nm and 90 degrees, respectively. The conditions of incident light are the same as the grating model. The result is shown in Fig. 4.42. In this case, an absorbance mode can be observed from (X, Y) = (1.6, 7.5) to (0, 9.2). Though this absorbance line is not identical to the theoretical diffraction mode, their origin seems to be the same at around (1.6, 7.5). We think the diffraction mode of the ideal sinc curve is suppressed to the lower side by varying the surface configuration to the nanogap structure. As a result, the absorbance occurs at about ω/c = 9.2, which corresponds to 680 nm when the incident angle is perpendicular to the sensor surface (k = 0). Furthermore, the suppression of the absorbance

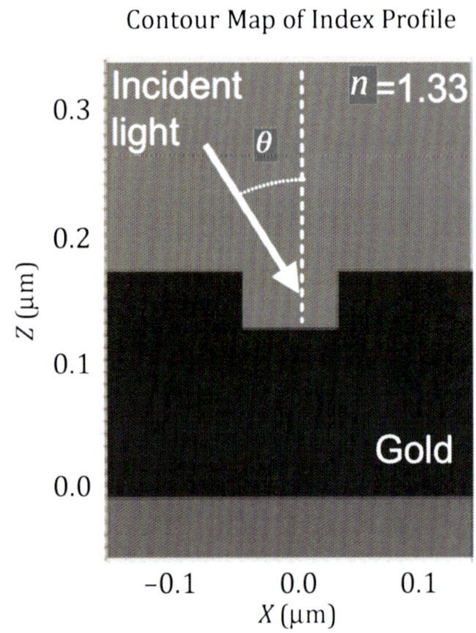

Figure 4.41 Simulation model for calculating the dispersion relation of the SPR mode on the nanopatterned surface.

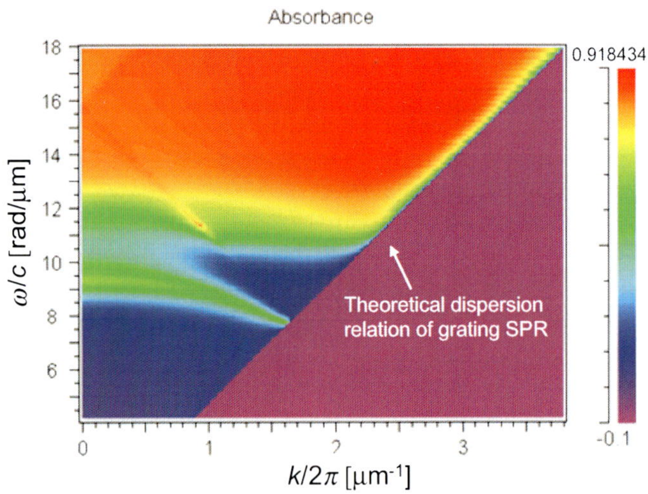

Figure 4.42 RCWA simulation result of the absorbance intensity (color contour) on a nanopatterned surface expressed as ω–k dispersion relation. The white line is the theoretical dispersion relation of the grating SPR, which is calculated based on the theoretical equations.

mode causes that the inclination of the curve at $k/2\pi$ is almost 0. Since the inclination of the dispersion relation curve corresponds to the group velocity of wave, the generated mode does not propagate along the sensor surface but stays inside the nanogap structure. These analyses revealed that the localized SPR on the nanoimprint pattern is the resonant mode modulated from the conventional grating SPR mode. Furthermore, the group velocity of the nanoimprint SPR mode is almost 0 and does not propagate along the sensor surface. This characteristic makes it possible for the generated electric field to be confined inside the nanogap structure and the sensing depth to be much smaller than the diffraction limit of light.

4.2 Fabrication of Sensor Device

The sensor device should be disposable for avoiding the blood contamination in order to be used as a practical biosensor. This means that very high reproducibility is demanded in the fabrication process of each biosensor. Furthermore, the device cost must be reasonable to be used extensively in various places. Localized SPR is expected as one of the most promising detection principle since it has the potential to provide easy, high S/N, small detection system. To generate localized SPR, various methods using metal colloid (Okamoto et al., 2000; Rechberger, 2003; Nath and Chilkoti, 2004), metal nanorod (Vial et al., 2007; Sepúlveda et al., 2009), polystyrene sphere (Haes et al., 2004; Kim et al., 2006), or electron beam lithography (Byun et al., 2006) are proposed and studied as introduced in Chapter 2. However, in the methods using small particles such as colloid, nanorod, and spheres, it is challenging to prepare those particles on the substrate with high reproducibility in their density since the density variations or aggregations sometimes occur in immobilizing them. In addition, the process throughput of the electron beam lithography drawing the nanopatterns for each device is too low to be used for the mass production of disposable products. And it is supposed to take more than a few hours for fabricating each device.

In the previous subsection, it is proved that the periodic nanogap structure with appropriate pattern dimensions on the metal surface can generate the SPR whose intensified electric field is localized in a few tens of nanometers. Based on this principle, we have fabricated a biosensor device by using nanoimprint technology to resolve the conventional problems.

4.2.1 Fabrication Process

The fabrication process flow which we have proposed is shown in Fig. 4.43. The detailed explanations for each process are described as follows:

(i) Preparation of master substrate

As the first step, the nanopatterns are drawn on the e-beam resist (ZEP520A, ZEONREX Electronic Chemicals) coated on a silicon wafer of 8 inch diameter by electron beam system (ELS-7000, ELIONIX, Japan). The thickness of the e-beam resist spun coated on a silicon wafer is a considerable parameter since its thickness determines the depth of nanogap structure. In this case, its thickness is adjusted to be 40–80 nm. And the dose parameter of electron beam is set as 0.20 eV. In this process, it takes about 5 hours to draw the nanopatterns with 280–300 nm pattern periods in about 45 mm square.

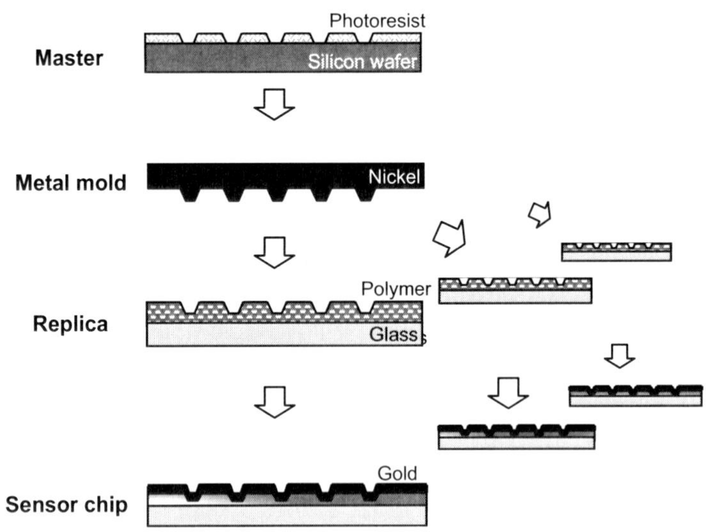

Figure 4.43 Schematic of the fabrication process flow of the nanoimprint biosensor.

(ii) Mold fabrication

As a next step, nickel mold is fabricated from the master substrate. To prepare the conduction layer for the following electroforming process, a thin nickel layer (thickness, 100–200 nm) is formed on the master substrate by sputtering process (CS-200S, ULVAC, Japan). The nanopatterns transferred on this nickel layer become the nanopatterns on the mold. Therefore, the thickness of this layer should be thicker than the depth of nanopatterns on the master substrate. Then the master substrate is immersed in the electroforming chemical liquid. And the mold layer (thickness, 250–300 μm) is formed by the electroforming process (SA1m, Digital Matrix, USA). This process takes about a few hours in general. The fabricated mold is separated from the master substrate and is cleaned by chemicals such as acetone and IPA (isopropyl alcohol) after the electroforming process. Since the resist pattern on the master substrate is destroyed in this process, it is difficult to use the master substrate repeatedly in this method.

(iii) Replication process

Subsequently, the nanopatterns on the mold are transferred onto the surface of the replica by using nanoimprint technique. The UV curable or thermoplastic polymer is generally used for this process. In this process, it is possible to produce one replica substrate in about from a few seconds to a few minutes. And the multiple devices can be fabricated from one replica since multiple device patterns can be aligned in one substrate. One mold can be used through more than ten thousand shots as long as it is not damaged by the contamination and so on in general. The reproducibility of the replicated nanopatterns is very high since many replicas can be produced from the same mold patterns.

(iv) Formation of gold layer

Finally, a gold layer is formed on the replicas surface by using the sputtering equipment (CS-200S, ULVAC, Japan). If the adhesion between the gold layer and polymer surface is weak, an adhesion layer (Cr, Ti, etc.) whose thickness is 2–3 nm should be added between them. Since many replicas can be treated in a large chamber at once, the process tact time for each substrate can be sufficiently small.

As presented here, high reproducibility and high throughput of the sensor device can be realized by using this process. And the cost of the device can be low since the mold can be repeatedly used many times. It is expected that these characteristics can overcome the drawbacks of the conventional localized SPR devices.

4.2.2 Fabrication Result

The fabrication results of biosensor substrate using the nanoimprinting process are presented in this subsection.

4.2.2.1 Device fabricated by photopolymerization

As the first example, the device is fabricated through photopolymerization replication process by using UV curable polymer. On a glass substrate (S3131, Matsunami Glass, Japan), HMDS (hexamethyldisilazane) is coated to improve the adhesion of polymer. And the polymer is dipped subsequently. After that, the nickel mold is put on it and the UV source of 10 mW intensity is irradiated from the transparent glass side for 240 sec. After peeling off the mold, a gold layer of 100 nm thickness is sputtered on it. The picture of the fabricated sensor device is shown in Fig. 4.44. At the center of the substrate, the nanopattern is formed in the area of 6 mm square. This area is clearly observed to be colored red as the result of absorption of the green light in the air. The scanning electron microscope (SEM) top view images of the master, mold, and replica are shown in Fig. 4.45. About the master substrate, the cross-sectional view is also included in it. As seen in this cross-sectional picture, the top edge of the nanogap structure has a slight curvature because of the effect of the electron beam diffusion during the exposure process. The surface topology of master and replica is measured by atomic force microscope (AFM) (Dimension 3100 Nanoscope IV, Digital Instruments, USA) with a super sharp probe (Nanotools Inc., Super sharp short-type length 300 nm, radius 2–3 nm) as shown in Fig. 4.46. The period and depth of fabricated nanogap pattern are 300 nm and 80 nm in this case. It can be observed that the transfer ratio in the depth direction is almost 100%. About the width direction, the transfer ratio is about 100–120%, which means that the width of replica gets slightly wider. This is assumed to be caused by the nanogap expansion during the peeling off process of mold. From these results, it is verified that the sensor device with almost identical nanopatterns with the master substrate can be produced in only a few minutes by using this process. This means that the process time and cost can be much reduced.

Figure 4.44 Picture of the nanoimprint biosensor fabricated by UV imprint process.

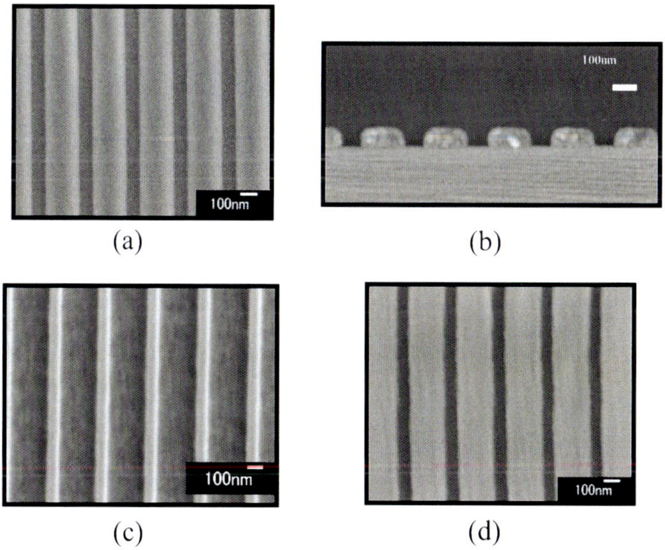

Figure 4.45 SEM images of (a) master substrate, (b) cross-sectional image of the master substrate, (c) mold substrate, and (d) UV replica substrate.

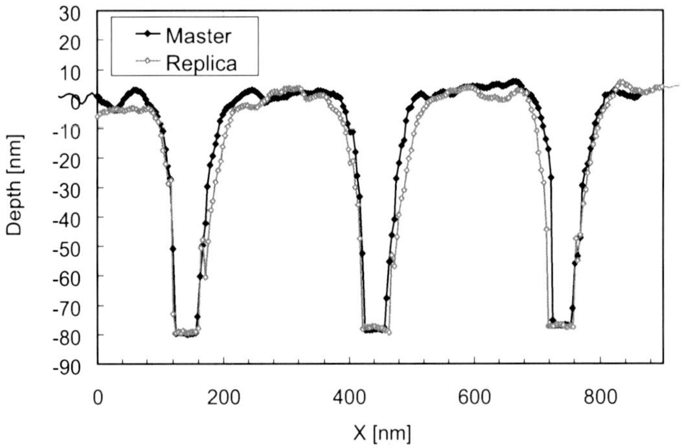

Figure 4.46 AFM measurement result of the master and replica surface.

4.2.2.2 Device fabricated by injection molding

As presented above, the nanopatterns can be accurately transferred on the surface of the replica through the photopolymerization process. While this process has the merit that the pattern accuracy is not affected by the heat cycle shrinkage, it requires the base glass substrate and it takes a few minutes. To reduce the material cost and to improve the process throughput more significantly, another approach using the injection molding is also performed.

In the injection molding, the thermoplastic resin heated near or over the glass transition temperature is injected into the mold with a high injection pressure (typically 10–3000 kgf/c) and released after cooling down. Though this process has been quite widely used for the fabrication of various products, the transcription of nanoscale patterns is very challenging issue. We have controlled the injection conditions to realize the nanoscale pattern transcription. The master and mold substrate are prepared as in the previous methods. The backside of the fabricated mold is polished to flat surface and is attached to the injection machine (J45AD-DK, Japan Steel Works Co., Japan) with other molds for injection.

As the replication polymer, polycarbonate (Panlite AD-5503, Teijin Kasei America Inc., USA) whose glass transition temperature is 144 degrees Celsius is used. The fabrication is carried out with the polymer heating temperature of 107, 112, and

117 degrees Celsius. Figure 4.47 shows the replicated product at 112 degrees Celsius. Two devices can be fabricated all at once. By sputtering 100 nm gold layer on it, seven nanopatterned arrays can be visible as shown in Fig. 4.48. The cross marks at the corner of the sensor device are also fabricated by nanopatterns and is used as the alignment mark at the spotting process of the probe liquid. In this experiment, as the adhesion between the polymer and the gold is enough, no adhesion layer is necessary. The surface topography images of the mold and the replicas measured by AFM (SPA-500, SII Nanotechnology Inc., Japan) are shown in Fig. 4.49. While the height of the nanopatterns on the mold is 33.9 nm, the heights of the replicated nanopatterns are 31.4 nm (107 degrees Celsius), 32.5 nm (112 degrees Celsius), and 32.1 nm (117 degrees Celsius), which means the transcription ratios are 93%, 96%, and 95%, respectively. Though the significant differences are not observed in terms of the replication temperature, it was verified that high precision transcription of more than 95% can be realized by raising the replication temperature up to 112 degrees Celsius. The width of the replicated nanopattern is about 100 nm which is smaller than the pit size of Blu-ray disc (150 nm). The cycle time of the injection molding in this case is 15 sec in which two devices are acquired. Therefore, the process time to fabricate one device is as small as 7.5 sec, which is much smaller than that of the photopolymerization process. And since this process does not require any glass substrate, the material cost can also be drastically low.

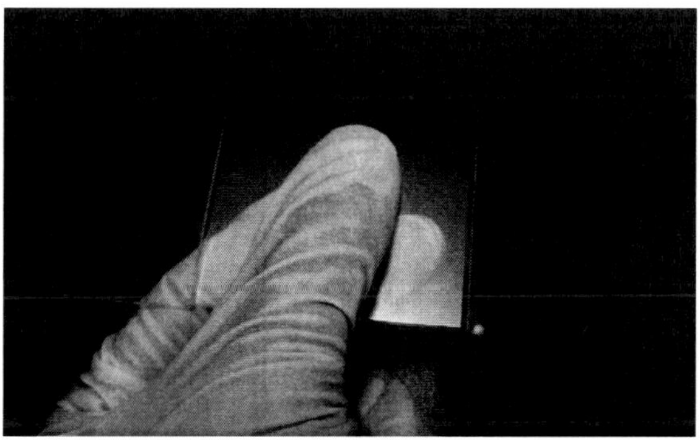

Figure 4.47 Picture of the replica fabricated by the injection molding process.

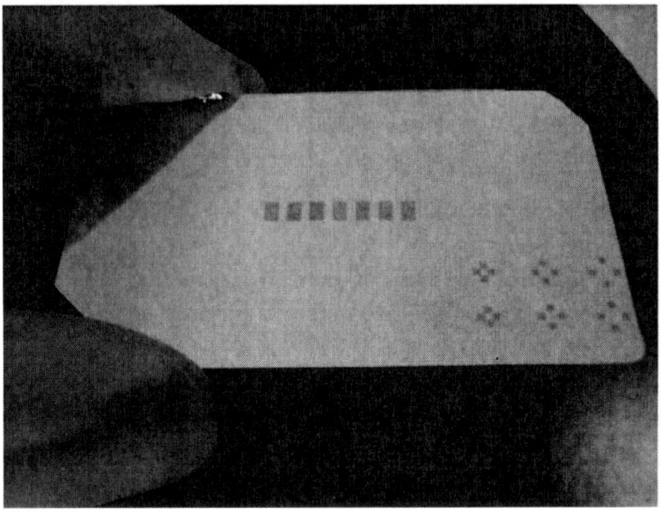

Figure 4.48 Picture of the replica after gold sputtering process.

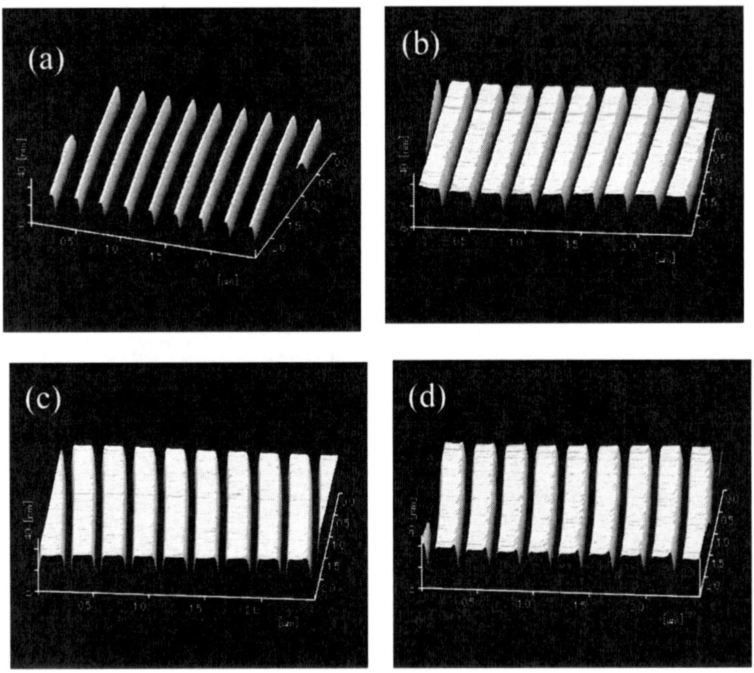

Figure 4.49 AFM images of (a) mold, replicas fabricated at (b) 107 degrees, (c) 112 degrees, and (d) 117 degrees Celsius.

4.2.3 Fabrication of Microchannel

4.2.3.1 Design of microchannel

To use the device for a biosensor, the microchannel which transports the liquid sample from the injection port to waste liquor tank through the sensing point plays a meaning role. Both depth and width of the microchannel are generally several tens of micrometers. And the liquid sample in the microchannel has the following unique properties compared with the bulk condition in a free space since the ratio of the surface area to the volume is getting larger. The liquid flow behaves as laminar flow and the control of the liquid sample is easy in the microchannel.

For using the biosensor, the volume of the sample should be as small as possible because the sample is very costly. Furthermore, the cross-sectional shape of the microchannel has a great effect on its detection sensitivity especially when the concentration of the target molecule is low. For example, if the reaction area is too long along the flow path, few target molecules can reach to the sensing area due to being captured before the sensing area. Also, if the depth of microchannel is too deep, the contact ratio of the target molecule on the reaction layer gets low. These phenomena are called mass transport limitation and its effect (MTC; mass transport coefficient) is expressed in the following equation:

$$\mathrm{MTC} = 0.98 \times \sqrt[3]{\frac{D^2 \cdot f}{h^2 \cdot w \cdot 0.3 \cdot l}}$$

where D is diffusion constant for the analyte; f is volumic flow; and h, w, and l are height, width, and length of the flow cell, respectively.

And the signal change rate per time can be expressed by the following equation:

$$\frac{\mathrm{d}R}{\mathrm{d}t} = \frac{k_\mathrm{a} \times C \times (R_{\max} - R) - k_\mathrm{d} \times R}{1 + \dfrac{k_\mathrm{a} \times G \times (R_{\max} - R)}{\mathrm{MTC}}}$$

where G is the response to surface concentration conversion factor, R is response, $R_{\max}$ is saturation response, k_a is association rate constant, k_d is dissociation rate constant, C is concentration of injected analyte, and t is the elapsed time after the injection of sample.

From this equation, we inferred that the larger MTC is, the more reaction efficiency is improved. To get higher signal, the depth of microchannel should be shallow and the width and length should be narrow and short and the flow rate should be fast.

As an example, the effect of the structural parameters on the signal is calculated by BiaSimulation (GE Healthcare, USA). The basic simulation parameters are shown in Table 4.9. The result assuming that AFP (α-fetoprotein, 70 kDa) of 100 ng/ml is sent through the various sizes of microchannel as shown in Fig. 4.50. Comparing our conventional microchannel (H0.2 mm × W1.5 mm × L11.5 mm) with that of Biacore-X system (H0.05 mm × W0.5 mm × L2.4 mm), the signal of latter can be about five times higher than the former. In this way, it is proved that the size of the microchannel is a very important element for the sensitivity of the sensor and should be small to get high sensitivity.

Table 4.9 Parameters for calculating mass transport coefficient

Parameter		Value	Unit
Sample volume	S	60	μL
Molecule weight	M_w	70000	Da
Association rate constant	k_a	4.70E + 05	M/sec
Dissociation rate constant	k_d	7.40E – 04	1/sec
Diffusion constant for the analyte	D	7.37E – 11	m^2/sec
Volumetric flow rate	f	20	μL/min
Height of flow cell	H	variable	
Width of flow cell	W	variable	
Length of flow cell	L	variable	

On the other hand, the pressure loss (ΔP) for the microchannel is represented in the following equation (Ducre and Zengerle, 2012):

$$\Delta P = 6.9 \cdot \mu \cdot f \cdot L \cdot \{(w+h)/(w \cdot h)\}^2$$

where μ is a viscosity coefficient, f is flow rate, L is the length, and w and h are width and depth of microchannel. The result on the condition of μ = 0.001 (water, 20 degrees Celsius), L = 40 mm, and f = 20 μL/min is shown in Fig. 4.51. This graph shows that the pressure loss extraordinarily increases when the size of microchannel gets smaller. As an example, it can be as high as 200 kPa at w = 0.5 mm, and h = 0.01 mm. In designing an actual sensor device, it is necessary to determine the size parameters in consideration of the pumping capability of flow pump.

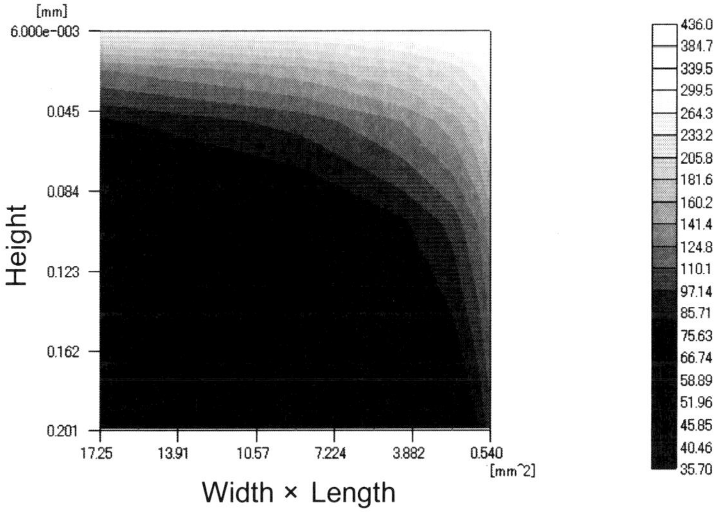

Figure 4.50 Calculated mass transport coefficient as the function of structural parameters of microchannel.

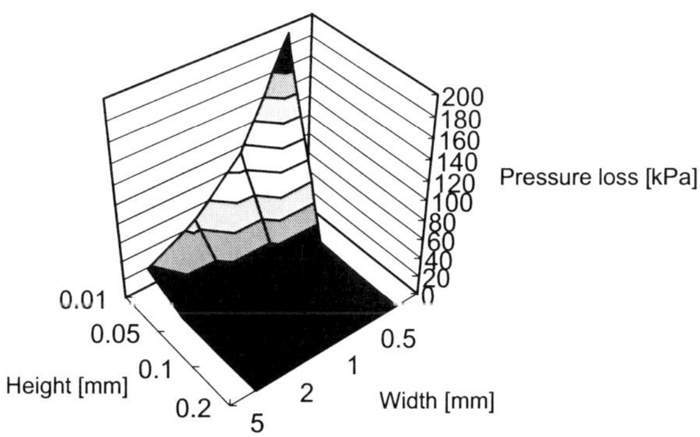

Figure 4.51 Calculated pressure loss under the conditions that the length of microchannel is 40 mm and the flow rate is 20 μL/min.

4.2.3.2 Microchannel made of PDMS

For the experimental use, the microchannel made of PDMS (polydimethylsiloxane) is fabricated. The schematic image and picture of the SUS mold prepared by cutting machine are represented in Fig. 4.52. The width and depth of the sensing area are 0.8 mm and 20 μm, respectively. Those of the other area are 2.0 mm and 200 μm, respectively. And the boundary of these areas is tapered. By enlarging the size of the microchannel other than the sensing position, the pressure loss can be reduced. Figure 4.53 shows the fabrication process of microchannel. PDMS (TSE3032(A):TSE3032(B) = 10:1, GE Toshiba Silicones Co., Ltd., Japan) was poured into the mold. And then, a glass substrate (Matsunami glass S3131, 28 mm × 48 mm, t = 0.8–1.0 mm) with the penetration holes at the positions of inlet and outlet was put on it and the polymer is cured at 100 degrees Celsius for 1 hour. The cured PDMS with the slide glass is peeled off from the mold and attached onto the sensor device. The microchannel is connected to the flow tube by using a commercially available connector (Nanoport™ Assemblies, Upchurch Scientific, USA) through the O-ring (CO 0000L, NOK Co., Japan). The PDMS microchannel is pressed on the sensor substrate by an original jig to prevent the liquid sample from leaking out. Figure 4.54 shows the pictures of the fabricated microchannel and sensor substrate. In this way, many microchannels can be produced by using mold and PDMS easily. And it can be used repeatedly by washing with ethanol. For the experimentally evaluation of sensor device in laboratory, this kind of microchannel is often used for its usability.

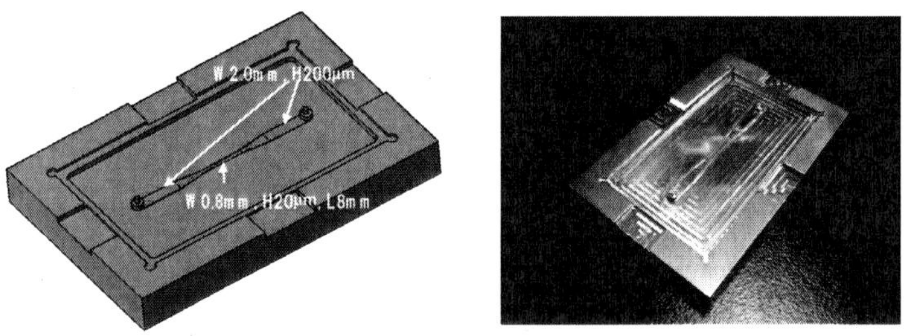

(a) (b)

Figure 4.52 (a) Schematic image and (b) picture of a mold for the PDMS microchannel.

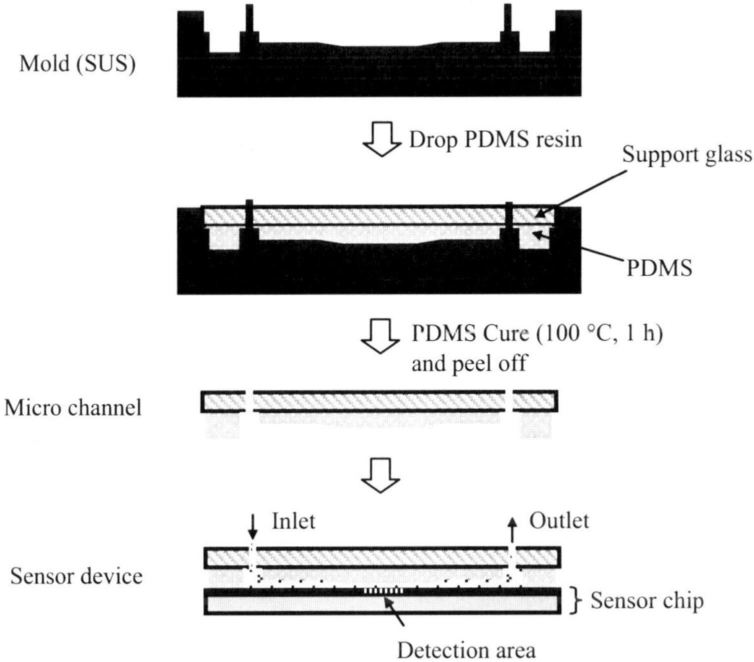

Figure 4.53 Fabrication process flow of the microchannel based on PDMS.

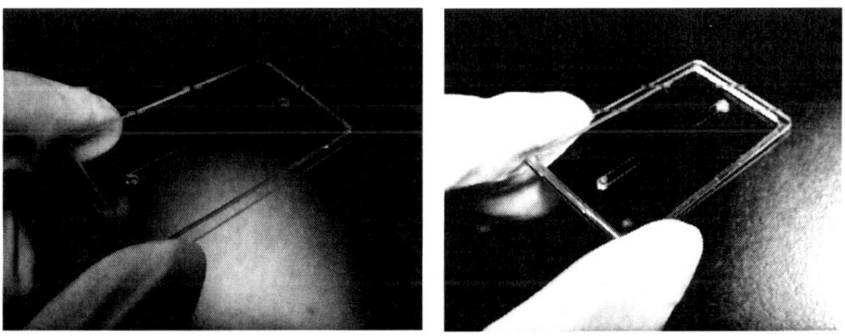

(a) (b)

Figure 4.54 Pictures of PDMS microchannel on (a) glass substrate and (b) set on the sensor chip.

4.2.3.3 Micro- and nanohybrid structure

As described above, the microchannel made of PDMS is easily used for the experiments. For the commercial applications, however, it takes too much time and labor to prepare it. To realize the mass production of microchannel and sensor device, we have proposed a hybrid fabrication process (Fig. 4.55) in which both the microchannel and the nanostructures are fabricated at once. Its basic concept is that the master structure of the microchannel is subsequently formed on the nanopatterned master substrate. Since it was found that the nanopatterns fabricated by the electron beam resist were damaged during the development process of microchannel photoresist, the modified nanopattern fabrication process is introduced here.

(i) Nanopattern fabrication for nanoimprint SPR by electron beam

Electron beam resist (ZEP520A, Zeon Co., Japan) is spun coated on a silicon wafer of 8 inch diameter. The thickness of the resist was 190 nm. And the nanopatterns are drawn on it by electron beam equipment (ELS-7000, Elionix, Japan). The exposure of electron beam penetrated through the resist completely. The resist was developed by o-xylene and washed by IPA and then dried by air blowing.

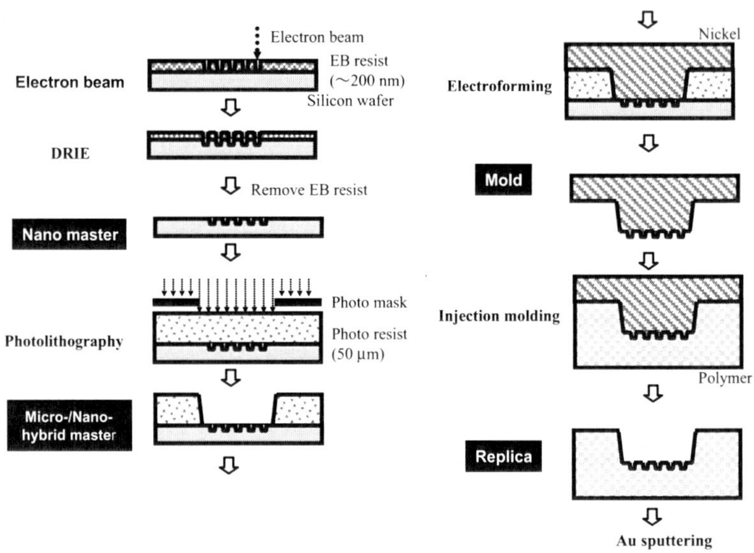

Figure 4.55 Schematic of process flow for micro-/nanohybrid structure.

(ii) DRIE (deep reactive-ion etching)

The master substrate after the electron beam process was dry etched by DRIE (RIE-200iPB). As the etching parameters,

Figure 4.56 Picture of micro-/nanohybrid chips fabricated by injection molding.

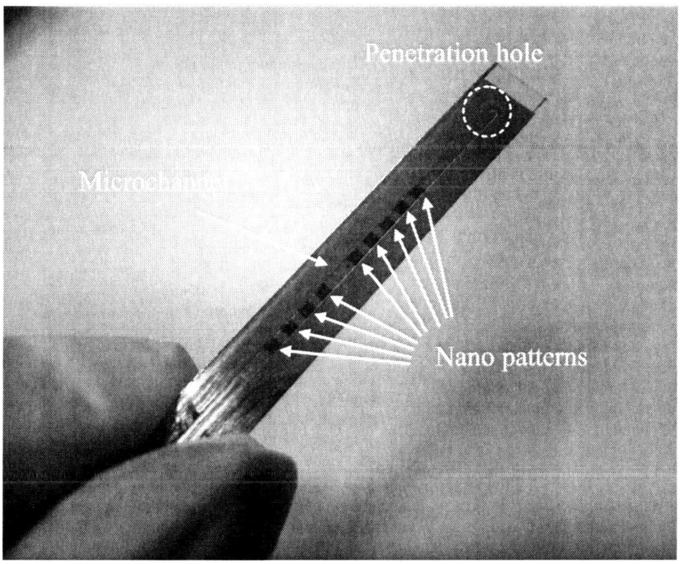

Figure 4.57 Picture of micro-/nanohybrid sensor device with microchannel and penetration hole.

the RF power was 160 W, etching time was 15 sec, and the etching gas was SF6. After that, the electron beam resist was removed by using a remover (ZDMAC, Zeon Co., Japan). By this process, the exposed bottom region of the electron beam resist was selectively dug on a silicon wafer. And the depth of the nanopatterns was adjusted to be about 40 nm on this fabrication condition.

(iii) Photolithography of microchannel

To prepare the microchannel structure, the photoresist (S1830, Shipley Company LLC, USA) was spun coated on the nanopatterned master substrate and its thickness was adjusted to 50 μm by controlling the spin speed. The photoresist was exposed by a photolithography equipment (FPA-3000iW, Canon, Japan) and developed after that to form the microchannel.

Based on these procedures, the master substrate of micro- and nanohybrid structure can be completed. Subsequently, the nickel mold is produced by electroform process in the same manner as described above. The replication is carried out by using injection molding process. Figure 4.56 shows the picture of the replicas. In this time, the outer shape of the sensor device is changed to a stick type (W6 mm × L48 mm × t0.85 mm) to reduce the useless area. The process tact time is about 15 sec per two pieces. This means that 10,000 pieces of sensor device with the microchannel can be produced within 1 day (about 21 hours). This is really high throughput process and has the potential to lower the process cost of sensor device.

The picture of the replicas with gold sputtering of 100 nm thickness is represented in Fig. 4.57. Since the outer shape is reduced this time, the sputtering process capability in one time goes up to 4.6 times more than the conventional device, which also leads to the shorter process tact time. From the picture, nine spots of nanopattern array (1.2 mm × 1.0 mm, 2 mm intervals) are observed inside the microchannel (W1.0 mm × H50 μm × L39 mm). While the penetration holes are necessary at the cover side in the conventional configuration, they can be formed at the base plate side during the injection molding process. This means that the nanopatterns and microchannel and penetration holes are fabricated all at once, which have been produced separately in the conventional process. Figure 4.58 shows the SEM images of the fabricated device. It is observed the nanogap structure of 70–80 nm width and 300 nm period is correctly aligned at the bottom of the microchannel whose depth is 50 μm. The nanopatterns in this study are changed to two-dimensional cross pattern structure in order to eliminate the polarization dependence while the nanopatterns are one-dimensional groove in the conventional study (the details will be described afterward).

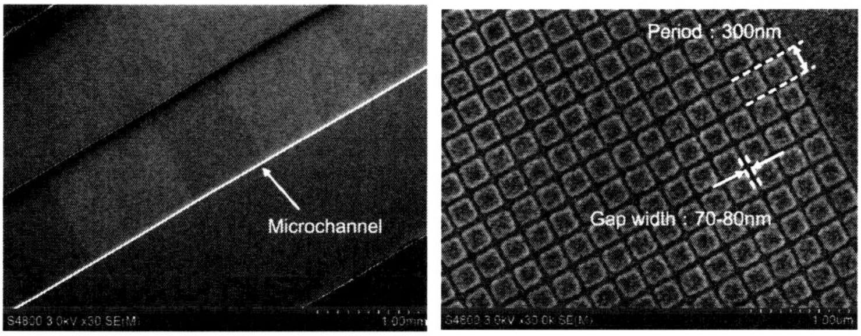

(a) (b)

Figure 4.58 SEM images of the hybrid sensor device: (a) microchannel and nanopattern area and (b) the magnified image of nanopatterned area.

As described above, we have proposed and implemented that the multistep fabrication processes of nanopatterns, microchannel, and penetration holes can be integrated to one process. These improvements can significantly lead to the cost down of the practical biosensors.

4.2.3.4 Sealing process of cover film

For streaming a liquid sample, it is necessary to seal the top side of the microchannel. As sealing technologies of microchannel, thermal compression bonding or the method using an ultraviolet curable resin is common. However, the microchannel can be deformed by heat and the probe proteins can be inactivated by the effect of heat or ultraviolet. In order to solve these problems, we have proposed a simple sealing method at room temperature that applies a transparent pressure-sensitive film (UC-500, 3M, Japan). Though this pressure-sensitive film had been originally developed and commercialized for sealing the sample plate of the biochemical tests such as ELISA (enzyme-linked immunosorbent assay), we found it is also very useful for sealing the microchannel. As shown in Fig. 4.59, the replicas are aligned on the jig and the pressure-sensitive film is put over them. Then the sealing is finished by applying the pressure on the backside of the film with a squeegee. In this method, no damages will occur on device and the sealing process is completed only in several tens of seconds without any special equipments. The surface profile of the microchannel (before and after sealing) and the sealing film (top and bottom side) are measured by using optical microscope as shown in Fig. 4.60. From this result, it is proved that the microchannel is not deformed by this sealing process and that the sealing film does not break into the microchannel and the depth of microchannel is kept as the designed dimension. In this case, the thickness of the film was about 70 μm. Figure 4.61 shows the sealed sensor device whose microchannel is filled with water. It is observed that the color of nanopatterned area is changed from red to green since the resonant wavelength changes. And it is also verified that there is no leakage of water.

Figure 4.59 Process picture of attaching a cover film on a sensor device.

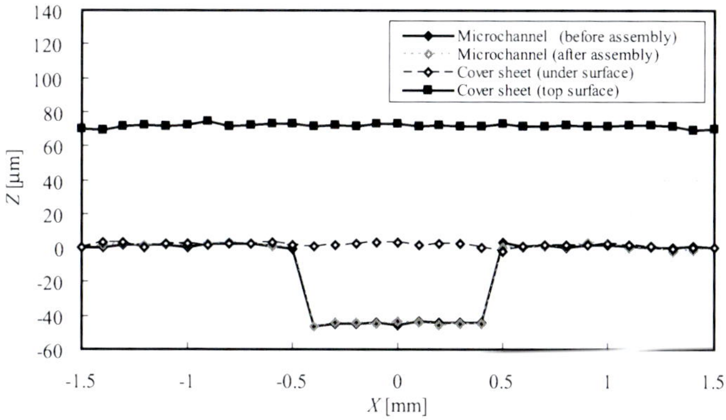

Figure 4.60 Surface profile data of the microchannel and cover film.

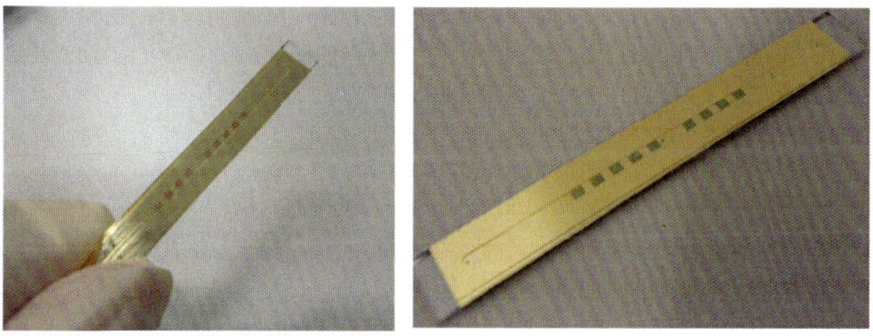

(a) (b)

Figure 4.61 Pictures of hybrid sensor device (a) in air and (b) with the microchannel filled with water.

As presented in this section, very simple and easy sealing of microchannel is realized by using the pressure sensitive film. By using this method, it is also possible to prepare the probe layer by dipping the device into the solution and the sealing is conducted afterward.

4.3 Evaluation of Nanoimprint SPR

As presented so far, it was proved that the nanopatterned SPR sensor can be fabricated by using nanoimprint technology. The optical properties of the fabricated sensor are basically significant to be used as a biosensor. In this section, the basic optical evaluation and some extra analyses are reported.

4.3.1 Basic Evaluation

4.3.1.1 Experimental setup

For the optical evaluation, the experimental setup is introduced in this section. The picture is shown in Fig. 4.62. As a light source, small-sized halogen lamp (LS1-LL, Ocean Optics Inc., USA) is used. The white light from the light source is collimated after eliminating the scattering elements through an aperture. And then, it can be linearly polarized through a polarizer. Subsequently, it enters into an objective lens (Plan N ×10, Olympus Co., Japan) after a beam splitter (BS CUBE NON-POL VIS 47121, Edmund Optics Inc., USA). The light is focused and exposed onto the surface of the sensor device. The actual spot size was 200–300 μm in this study. The reflection light from the sensor surface goes back to upward through the objective lens and is divided by the beam splitter. The divided light is collected by a small-sized spectrometer (USB4000, Ocean Optics Inc., USA). The reflection spectrum is analyzed on the PC.

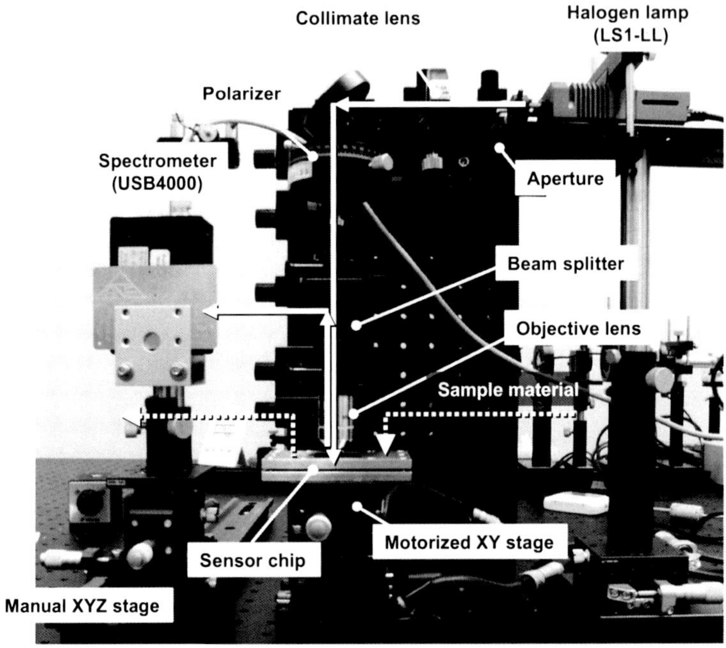

Figure 4.62 Picture of experimental setup for nanoimprint biosensor.

The buffer solution is sent to the sensor device by using a syringe pump (PHD2000, Harvard Apparatus Inc., USA). To insert a liquid sample solution, a manual injector (Manual sample injector 9725i, Rheodyne, USA) is applied on the way from the pump to the sensor device.

A motorized linear XY stage (SGSP15-10, Sigma Koki Co., Ltd., Japan) is, furthermore, prepared under the stage of the sensor device to control the position of sensor chip with the stepping motor controller (SHOT204MS, Sigma Koki Co., Ltd., Japan). The spectrometer is also set on the manual controllable XYZ stage to adjust its position. The software is constructed by using LabVIEW (National Instruments Corp., USA). And this system makes it possible to measure the multiple arrays and analyze the distribution errors of sensor chip.

4.3.1.2 Basic evaluation of nanoimprint SPR sensor

By using the experimental setup mentioned above, the basic evaluations of the fabricated nanoimprint SPR sensor is carried out. The device was fabricated through photopolymerization process described in the previous subsection. In the first instance, the detailed nanopatterned surfaces of the replicas before and after the gold sputtering were measured by AFM (Dimension3100, Digital Instruments, USA). The cross-sectional dimensions based on the AFM results are drawn in Fig. 4.63. The width sizes are calibrated with considering the top radius (R < 5 nm) of the measuring probe (SS-ISC, Team Nanotec GmBH, Germany). From this result, the following shape transformations were observed after the gold sputtering. The bottom gets narrower since the gold layer is formed on the sidewall of the bottom. On the other hand, its top gets wider since the top edge gets rounded by gold layer formation. The depth of the nanogap gets deeper since the sputtering formation rate at the bottom gets lower than the top surface by shading effect. And as the aspect ratio of nanopatterns is larger, this effect becomes large.

The reflection spectrum obtained by using this sensor device is shown in Fig. 4.64(a). As the sample material, air (n = 1.00), water (n = 1.33), and ethanol (n = 1.36) are sequentially sent over the sensor surface. The reflection dips were observed at the wavelength of about 550 nm, 650 nm, and 660 nm when the sample material is air, water, and ethanol, respectively. These dips are caused by the absorption of the incident light since the energy of incident electric field is transferred to the oscillation of the free electrons in the gold layer. This result shows that the fabricated nanoimprint device can actually work as a sensor that detects the change of the refractive index of sample layer.

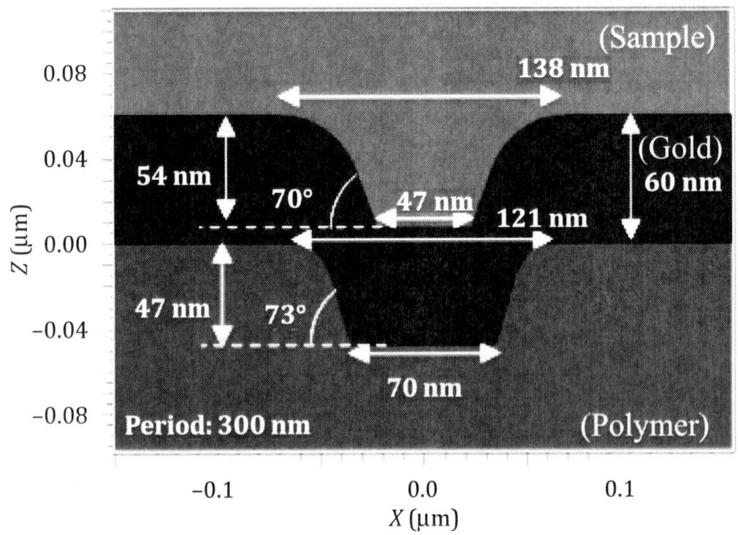

Figure 4.63 Cross-sectional detailed dimensions of the fabricated nanopatterns on the sensor device measured by AFM.

To compare this result with the simulation method, the reflection spectrum is calculated by using RCWA method. The simulation model is designed based on the measurement result as shown in Fig. 4.63. Figure 4.64(b) shows its result. The refractive index of the sample layer is changed from 1.00 to 1.33 and to 1.36 as is in the experiment. Table 4.10 shows the dip wavelengths obtained by the experiment and simulation and the sensitivity. The sensitivity is calculated by dividing the wavelength shift amount by the change in refractive index (1.36–1.33 = 0.03). Furthermore, the sensing depth is simulated by using the technique as described in Section 4.1. The result is shown in Fig. 4.65. According to the fitted curve, the sensing depth defined as the distance where the sensitivity decreases to $1/e$ times that of the sensor surface is about 30 nm. This result

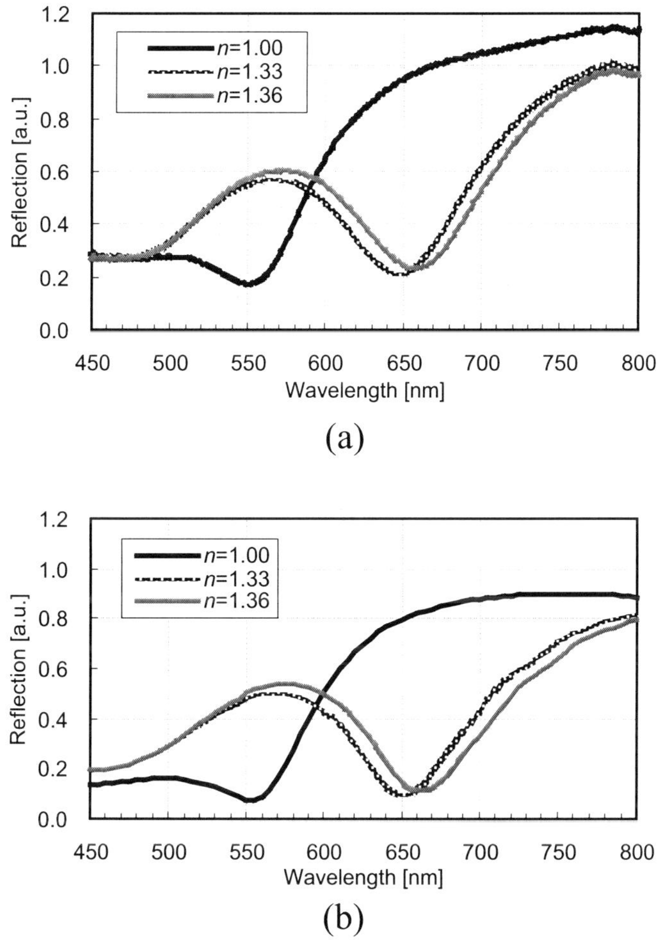

Figure 4.64 Reflection spectrum derived from (a) experiment and (b) simulation.

means that the localization of sensing region will be achieved by this nanoimprint device, though the edge of nanopatterns is not so sharp.

The bulk sensitivity is also compared with that of the conventional localized SPR using gold colloid. The gold colloids of 40 nm diameter (EMGC40, British Biocell International, Ltd., Britain) are fixed on a glass substrate (Fig. 4.66). And the resonant wavelength is evaluated by observing the absorbance peak on the conditions that the sample is air, water, and ethanol. The results are shown in Fig. 4.67. The bulk sensitivities (from air to water) were calculated from the inclination of the fitted curve. The bulk sensitivity of the nanoimprint SPR sensor is 305 nm/RIU, while that of the conventional colloid SPR sensor is 134 nm/RIU. This means that the bulk sensitivity of nanoimprint SPR can be two times higher than that of the conventional LSPR sensor. This seems to be caused by the generated intensified electric field that is strongly confined and enhanced inside the nanogap structure than the sphere structure whose enhanced electric field is formed outside. According to these results, the nanoimprint SPR sensor has the advantages not only in its production and wavelength tuning aspects but also in the sensitivity.

Table 4.10 Resonant wavelengths and calculated sensitivity of experiment and simulation results

	Experiment	Simulation
Peak λ @ n = 1.00 (air)	547.1	548.5
Peak λ @ n = 1.33 (water)	647.0	651.3
Peak λ @ n = 1.36 (ethano)	657.3	663.0
Sensitivity [nm/R U]	343.5	390.1

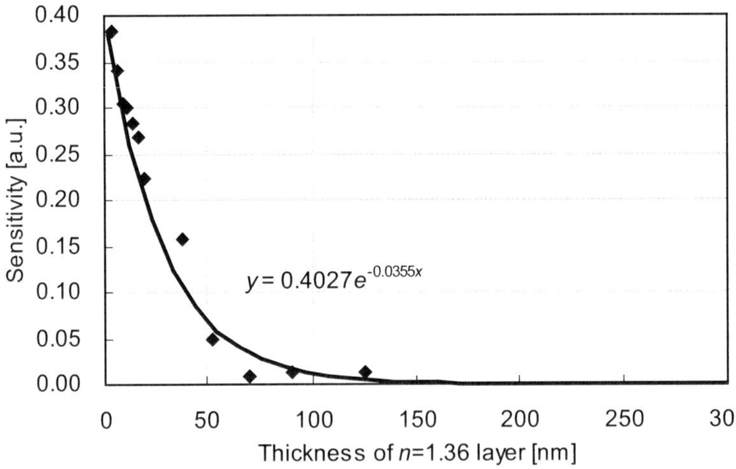

Figure 4.65 Calculated sensitivity as a function of the thickness of n = 1.36 layer attached on the gold surface.

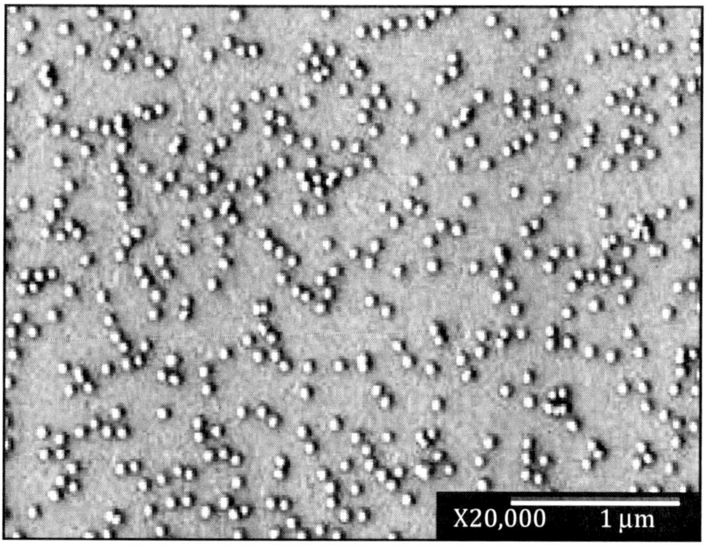

Figure 4.66 SEM image of the gold colloids immobilized on the glass substrate.

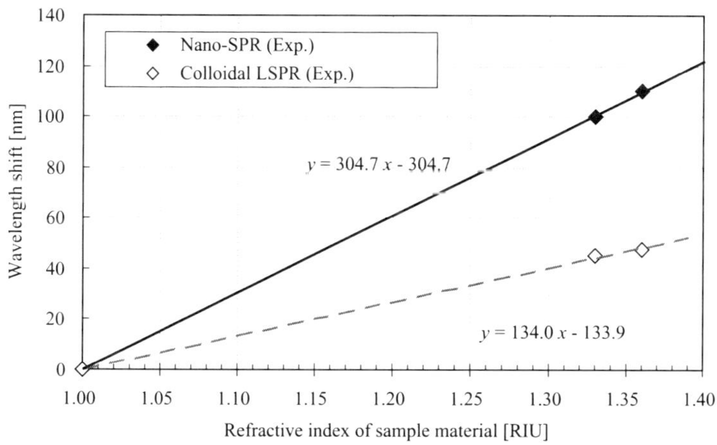

Figure 4.67 Comparison of the bulk sensitivity between the gold colloidal LSPR and nanoimprint LSPR sensor.

4.3.1.3 Experimental evaluation of sensing depth

In this section, the sensing depth of the fabricated device is evaluated experimentally by adding SiO_2 layer on it. The chromium layer of

2–3 nm thickness is prepared between the gold and SiO_2 layer as an adhesion layer. And the SiO_2 layers of different thickness are sputtered on the sensor surface (CS-200S, ULVAC, Japan). These two sputtering processes are sequentially conducted without venting the sputtering chamber. The typical refractive index of the SiO_2 is 1.46.

Based on the above process, some sensor devices with different SiO_2 thickness are prepared. On the sensor surface, two different solutions were funneled as a sample material. And the signal shift between these two sample materials was observed (Fig. 4.68). The first solution is ethanol (n = 1.36) and the second solution is the mixture of glycerin and water whose refractive index is adjusted as 1.46 (volume ratio of glycerin to water is 93% to 7%). When the sensing depth is deeper (larger) than the thickness of SiO_2 layer, the shift of resonant wavelength will be observed due to the change of the refractive index of the sample material. On the other hand, when the sensing depth is smaller than the thickness of SiO_2 layer, no signal shift will be observed since the sensor will not detect the change in the sample material. Based on this principle, the sensing depth of the sensor device is evaluated. As comparisons, the sensing depth of the conventional propagating SPR and the conventional colloidal LSPR is investigated. As the propagating SPR device, the sensor device with a gold layer of 50 nm thickness is used. And the total internal reflection spectrum is observed by attaching an optical prism with matching oil. As the colloidal LSPR device, the previously introduced substrate with 40 nm gold colloids is used. And the absorbance spectrum is observed. The chromium layer is also prepared as an adhesion layer for SiO_2 in both cases.

The result is shown in Fig. 4.69. In the case of the propagating SPR, the signal shift gradually decreases as the thickness of SiO_2 layer increases. In the case of nanoimprint SPR sensor, the signal shift sharply deceases as the SiO_2 thickness increases. The sensing depths (the distance where the signal shift gets $1/e$ times smaller) derived from the fitted curve for propagating SPR, nanoimprint SPR, and colloidal LSPR are 294 nm, 98 nm, and 123 nm, respectively. By this result, it is experimentally proved that the sensing depth of the nanoimprint SPR sensor is confined under 100 nm from the sensor surface. This means that the sensing depth of nanoimprint SPR can overcome the diffraction limit of light as the conventional LSPR. The reasons why the experimental result is larger than the simulation method in the previous section are assumed due to the effects of the chromium layer and the penetration of the liquid solution into the SiO_2 surface.

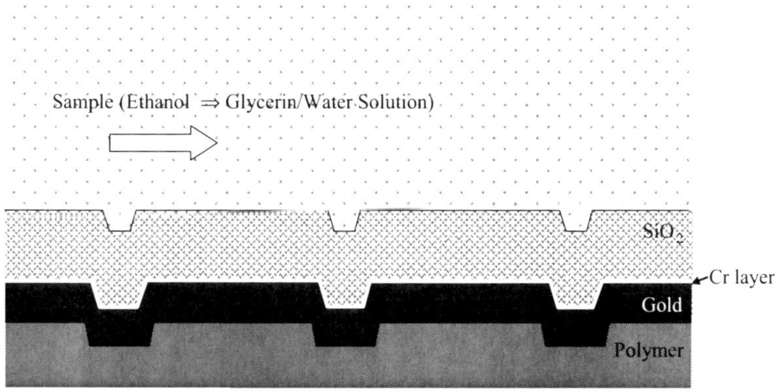

Figure 4.68 Schematic of the experimental model to evaluate the sensing depth of SPR sensor.

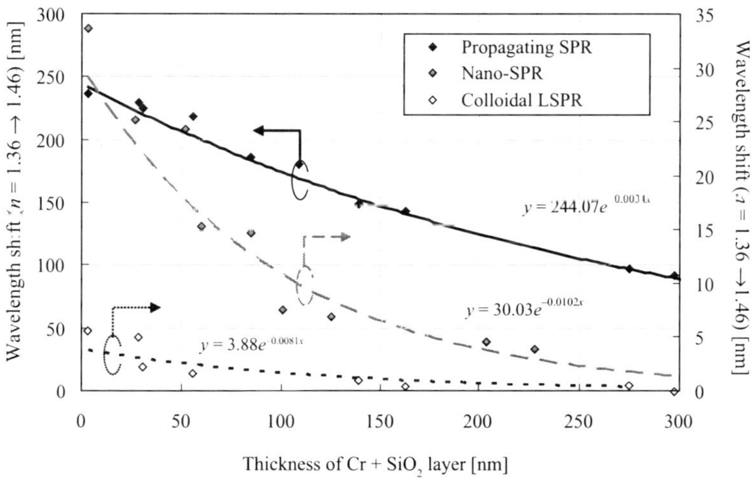

Figure 4.69 Experimental result of the sensing depth evaluated on the propagating SPR, nanoimprint LSPR, and gold colloidal LSPR sensor devices.

4.3.1.4 Experimental evaluation of sensing depth II

In this section, the effect of the localization of the sensing depth is analyzed by using real proteins for testing as a practical biosensor.

As the first step, protein A-fused protein (Orla18, Orla Protein Technologies Inc., Britain) is formed on the gold surface. And then, the sample solution with the anti-AFP antibody (0.03 mg/mL) is injected over the sensor surface as a target molecule. After that the concentration solution of FBS (fetal bovine serum, Tissue Culture Biologicals, USA) is used to test background noise. The sample volume is 100 μL and the flow rate is 40 μL/min. The same nanoimprint SPR device in the previous section is used in this experiment. As a comparison, the same experiment is carried out on the conventional propagating SPR using commercially available equipment (Biacore X, GE Healthcare, USA).

The experimental result normalized by the signal change of the anti-AFP antibody binding is shown in Fig. 4.70. "S" and "N" in the graph indicate the injection of anti-AFP antibody and FBS, respectively. In both cases, the nonreversible signal change is observed at "S" injection. This means that anti-AFP antibody binds to the prepared protein-A fused protein layer. In the propagating SPR, the signal change which is about four times as large as that of "S" injection is observed at "N" injection. It is also observed that this signal falls down to the point before "N" injection after about 2.5 min from "N" injection. This result means that most of the injected FBS does not bind to the sensor surface and that the signal change after "N" injection is caused by the floating FBS molecules over the sensor surface. Due to the floating FBS, the average refractive index in the sensing depth increases and the signal change is observed.

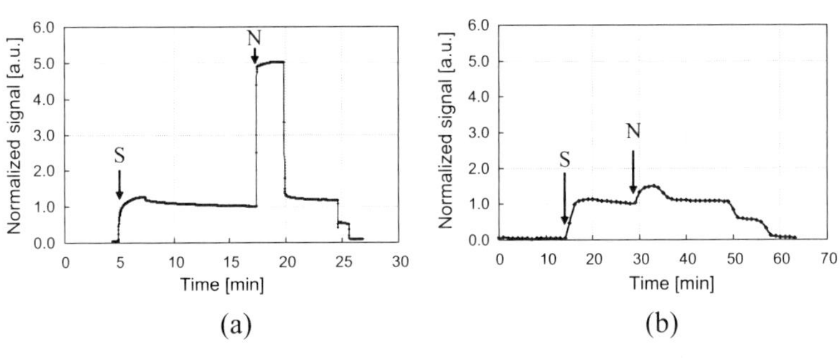

Figure 4.70 Experimental results of bulk effect evaluation using FBS on (a) propagating SPR and (b) nanoimprint SPR sensor. "S" is the injection of anti-AFP antibody which specifically binds on the sensor surface and "N" is the injection of FBS which only floats over the sensor surface.

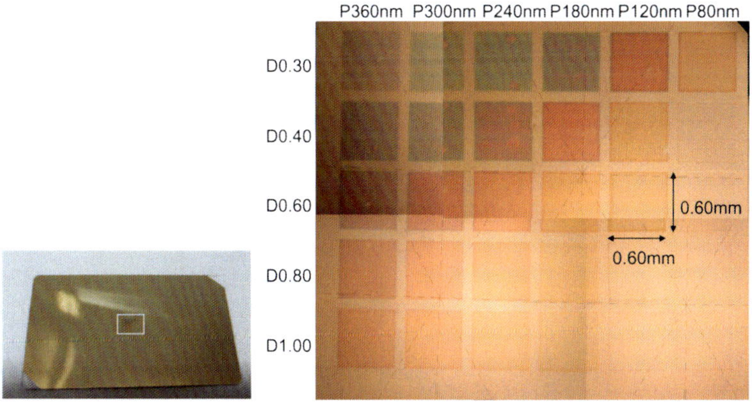

(a) (b)

Figure 4.71 (a) Pictures of the fabricated device on which the matrix pattern is prepared and (b) matrix pattern magnified by microscope.

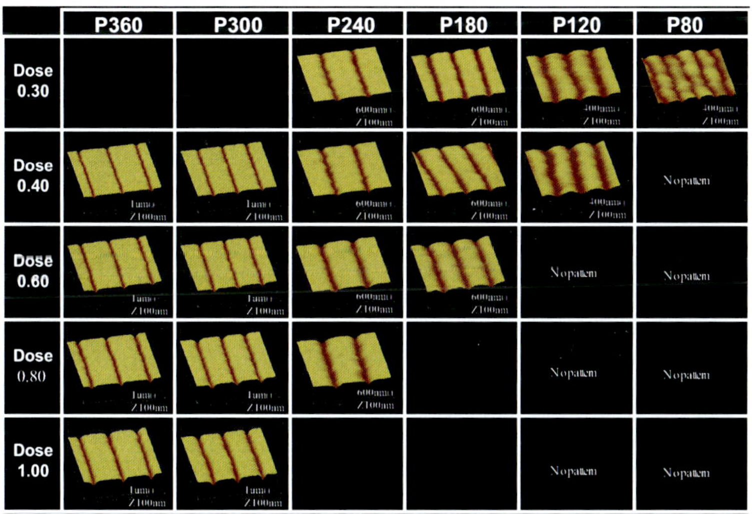

Figure 4.72 AFM images of the nanopatterns prepared on the matrix pattern.

On the other hand, the signal change by "N" injection is about 0.4 times as large as that by "S" injection. This signal also seems to be caused by the floating FBS molecules since this signal change falls down after about 2.5 min from "N" injection. In this case, it is verified that the signal-to-noise ratio of the nanoimprint SPR sensor can be about ten times higher than that of the conventional propagating SPR. This means that the nanoimprint SPR sensor is less sensitive to the background noise than the propagating SPR. In the conventional SPR equipment, the temperature of the sample solution is rigorously controlled (generally, 0.01 degrees Celsius) to reduce the background noise. And the wash out of the floating molecules is necessary to eliminate the background noise. In the nanoimprint SPR device, these complex control systems might be omissible with the effect of the localization of sensing depth. This has the potential to realize a low-cost and small-sized, high-sensitive biosensor.

4.3.2 Effect of Structural Parameters

4.3.2.1 The effect on the resonant wavelength

Next, we have experimentally evaluated the effect of the structural parameters of nanopattern on the optical properties. An e-beam resist of 46 nm thickness is prepared and the different nanopattern arrays are fabricated by adjusting the electron beam parameters. The periods of nanopattern are 80 nm, 120 nm, 180 nm, 240 nm, and 300 nm. And the exposure doses energy of electron beam which dominate the width of nanogap are 0.30 eV, 0.40 eV, 0.60 eV, 0.80 eV, and 1.00 eV. The size of each pattern area was 0.6 mm squares. And these patterns are aligned as matrix array. Subsequently, mold fabrication and replication (photopolymerization) processes are conducted as described above. After that the gold layer of 60 nm thickness is sputtered on it. The pictures of an external appearance and the microscopic image are shown in Fig. 4.71.

The AFM images of each pattern are shown in Fig. 4.72. It is apparently observed that the width of nanogap gets wider by increasing dose energy. And the top edge curvature of nanogap collapsed at the small period patterns (120 nm and 80 nm) due to the exposure interference effect by the neighboring nanogap. At the arrays with small pattern period and high dose energy, the nanopatterns are not formed because the e-beam resist is completely exposed in this region. About the depth of nanogap, the AFM probe could not reach the bottom of nanopatterns at patterns with low dose energy since the width of nanogap is too narrow. Basically, the depth should correspond to the thickness of e-beam resist. However, the actual depth of the patterns where the pattern period is small seems to be smaller than the thickness of resist because the resist other than the groove position is also exposed by e-beam leakage exposure. The patterns in these regions are supposed to be shallower than the other region.

With respect to the optical properties, it is observed that the color of pattern shift from blue to red by increasing the dose energy, that is, by increasing the width of nanogap is shown in Fig. 4.71. And the same color shift is observed by decreasing the pattern period. These results mean that the optical absorption range shifts from red to blue by increasing dose energy or by decreasing the pattern period.

The reflection spectra obtained by using these nanopattern arrays are shown in Fig. 4.73. Water and ethanol are used as the

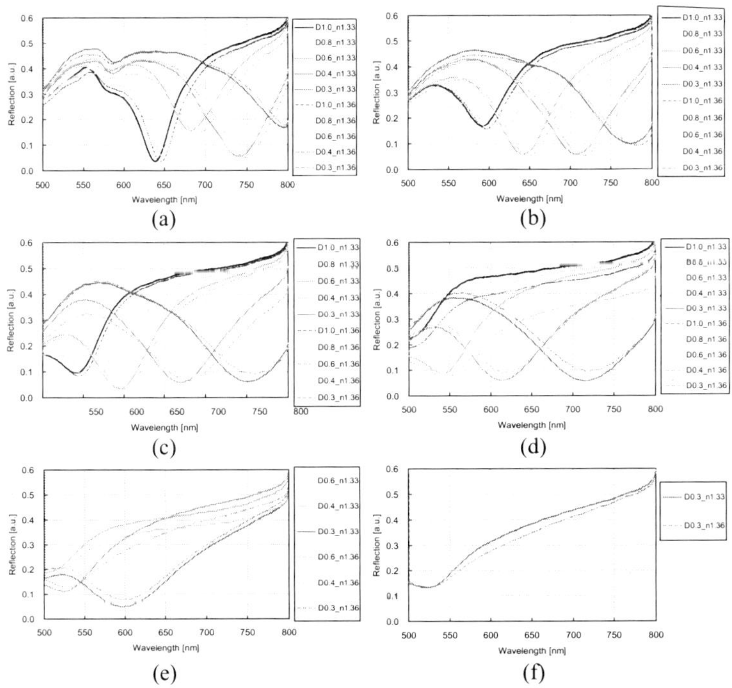

Figure 4.73 Reflection spectra of the matrix pattern. Each pattern period is (a) 360 nm, (b) 300 nm, (c) 240 nm, (d) 180 nm, (e) 120 nm, and (f) 80 nm.

sample materials with the flow control system. The resonant wavelength shifts to shorter wavelength when the dose is larger or when pattern period is smaller. These results correspond to the above supposition. The resonant wavelength in this experiment is depicted in Table 4.11. Also, the tendency of these results is correctly identical to the simulation method introduced in Section 4.1.

Table 4.11 Resonant wavelength and sensitivity of the matrix pattern

	Pitch-Dose	**P360-D1.0**	**P360-D0.8**	**P360-D0.6**	**P360-D0.4**	**P360-D0.3**
	λ @ n = 1.33 [nm]	638.5	652.4	682.6	740.8	–
	λ @ n = 1.36 [nm]	645.6	660.1	691.1	747.9	–
P360	**Sensitivity [nm/RIU]**	**237.5**	**255.5**	**282.3**	**236.8**	
	Pitch-Dose	**P300-D1.0**	**P300-D0.8**	**P300-D0.6**	**P300-D0.4**	**P300-D0.3**
	λ @ n = 1.33 [nm]	591.2	609.4	642.0	707.8	773.5
	λ @ n = 1.36 [nm]	596.4	615.8	648.8	715.6	791.3
P300	**Sensitivity [nm/RIU]**	**172.8**	**213.8**	**224.6**	**259.6**	**594.8**
	Pitch-Dose	**P240-D1.0**	**P240-D0.8**	**P240-D0.6**	**P240-D0.4**	**P240-D0.3**
	λ @ n = 1.33 [nm]	539.9	560.3	594.4	670.3	751.4
	λ @ n = 1.36 [nm]	543.5	564.3	599.0	679.1	758.1
P240	**Sensitivity [nm/RIU]**	**120.6**	**136.1**	**154.1**	**292.4**	**221.3**
	Pitch-Dose	**P180-D1.0**	**P180-D0.8**	**P180-D0.6**	**P180-D0.4**	**P180-D0.3**
	λ @ n = 1.33 [nm]	–	–	540.0	614.0	714.7
	λ @ n = 1.36 [nm]	–	–	543.7	618.5	720.3
P180	**Sensitivity [nm/RIU]**			**123.5**	**150.1**	**187.8**

4.3.2.2 The effect on signal-to-noise ratio

The effect of the structural parameters on the signal-to-noise (background noise) ratio is experimentally evaluated in this section. The same kind of matrix arrayed nanoimprint device introduced in the previous section is used with 100 nm gold sputtering. In Section 4.3.1, the signal-to-noise ratio is analyzed by using the specific binding of antibody on the surface (signal) and the floating FBS molecules (background noise). More unsophisticated method is adopted to simplify the experiment here. As a signal response, nonspecific binding of bovine serum albumin (BSA), Sigma Chemical Company, USA, is used. The BSA solution of 0.5 mg/mL is injected with the flow rate of 20 μL/min. After that the buffer solution is changed from water to ethanol as a background noise response. Each signal (S) and noise (N) responses on the matrix patterned arrays and the calculated S/N are shown in Fig. 4.74. In the arrays of the same pattern period, the signal response by BSA binding gets higher at the smaller dose energy (narrow nanogap width). In the simulation analysis, the sensing depth of narrower nanogap width is more sharply localized. And this seems to lead the increase of signal by the binding of BSA on the surface. As a result, the S/N becomes higher at the smaller dose energy.

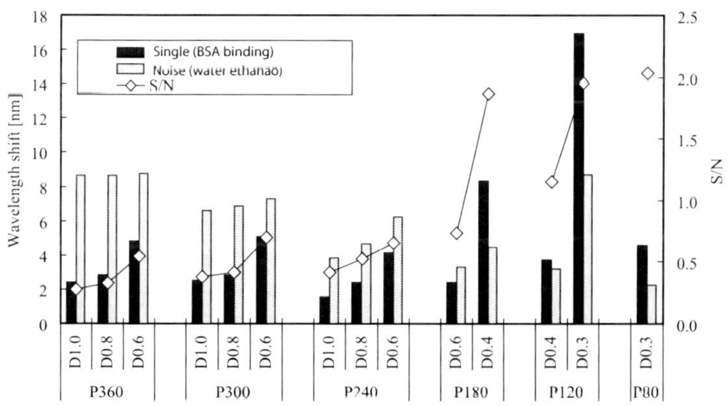

Figure 4.74 Signal-to-noise experimental results using FBS protein on matrix pattern.

Also, in the case where the pattern period is smaller, the S/N is observed to be improved. Especially, the S/N at the pattern period of 120 nm and 80 nm is two to three times higher than S/N at the pattern period of 300 nm. Considering the S/N evaluation result using protein-A fused protein and FBS in Section 4.3.1, the S/N evaluation at the pattern period of 120 nm and 80 nm can be 20–30 times higher than S/N evaluation of the conventional propagating SPR. These results verified that the sensing depth is more localized in the smaller pattern period, which corresponds to the simulation result. In this experiment, the probe layer is ideal since the direct binding of BSA on the sensor surface is treated as signal reaction. As described above, the nanopattern structural parameters should be designed with considering the thickness of the probe layer in practice for optimizing the signal-to-noise ratio. The experimental results in this section have verified that the signal-to-noise ratio of the nanoimprint SPR biosensor has the potential to improve its signal-to-noise ratio up to 20–30 times higher than that of the conventional propagating SPR.

4.3.3 Meshy (Two-Dimensional) Groove Pattern

4.3.3.1 Concept and principle of meshy groove pattern

The one-dimensional groove structures are basically analyzed in simulation so far. In this structure, however, only the light linearly polarized perpendicularly to the groove line can contribute to the excitation of localized SPR. This means that the incident light that is not linearly polarized perpendicularly to the groove line can be the background noise element. So, it is necessary to prepare the polarization filter that depresses the utilization efficiency of the incident light and increases the device cost in the optical system. Furthermore, when there is a polarization defect in the sensor device itself, the polarization error cannot be eliminated by outer polarization filter.

To resolve these problems, the nanopattern structure whose localized SPR generation is independent on the polarization direction of the incident light is studied. Though the simulation tools of FDTD method and RCWA method are applicable to the three-dimensional simulation, much more simulation memory and time is, in practical, necessary when compared with the two-dimensional simulation. Through our studies, it was found that the accurate simulation analysis is still challenging even by using several computers in parallel. On the other hand, the resonant behavior can be separately studied in each polarization direction. So, it is theoretically assumed that the meshy groove structure can provide the polarization-independent SPR device. The actual meshy groove structure is prepared on the sensor device and evaluated in this section.

4.3.3.2 Device fabrication and optical evaluation

The master pattern of the meshy groove structure can be fabricated by the electron beam process in which the meshy exposure lines are programmed. After that the mold is fabricated by the electroforming process. And the replicas are prepared by photopolymerization process. The gold layer of 100 nm thickness is sputtered on it. The AFM images of the fabricated devices with the conventional groove structure and meshy groove structure are shown in Fig. 4.75. The pattern period and depth are 300 nm and 43 nm in both cases. The pattern widths measured at the 10 nm down position from the top surface are 68 nm in the conventional pattern and 82 nm in the meshy groove pattern. The reason why the pattern width of the meshy groove pattern is wider than that of the conventional groove pattern is that the total amount of the electron beam exposure in the meshy groove pattern is larger than that of the conventional groove pattern.

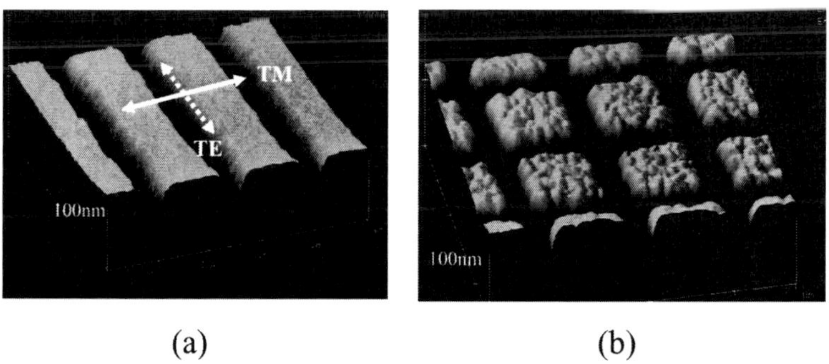

Figure 4.75 AFM images of (a) one-dimensional groove structure and (b) two-dimensional (meshy) groove structure.

The optical properties of these sensor devices are evaluated. About the polarization direction in the one-dimensional groove structure, the polarization which is perpendicular to the groove line is defined as TM and the polarization which is parallel to the groove line is defined as TE. The experimental results are shown in Fig. 4.76. About the conventional groove structure, the dip of the reflection spectrum is observed at about 620 nm in TM polarization condition. The shift of the resonant wavelength is also observed when the sample material is changed from water to ethanol. On the other hand, no apparent dip is observed in TE polarization condition. This result proves that the polarization component which is parallel to the groove structure does not contribute to generate the SPR. About the meshy groove structure, the dip of the reflection spectrum is observed at about 590 nm in both polarization components. The resonant wavelength of the one-dimensional groove structure is slightly larger than that of the meshy groove structure since the nanogap width of line groove structure is narrower than that of the meshy groove structure as examined above. When the randomly polarized light is used, the minimum value of the reflection spectrum in the one-dimensional groove structure does not reach

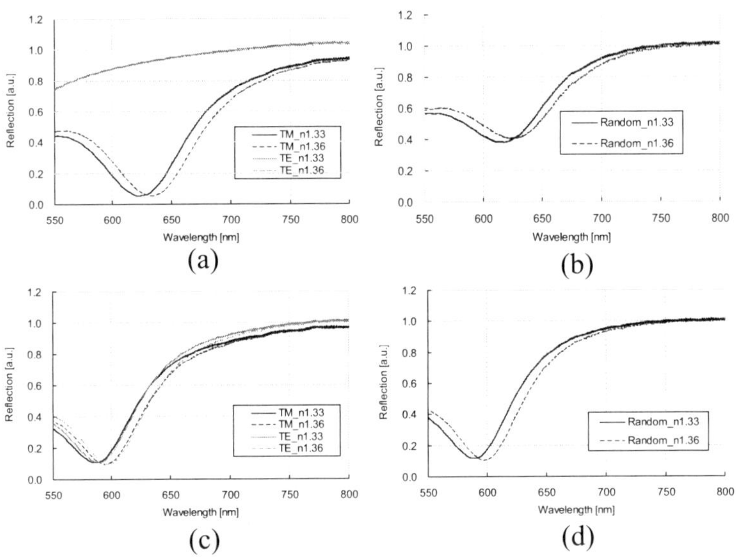

Figure 4.76 Experimental reflection spectra of groove structure with (a) linearly polarized incident light and (b) randomly polarized incident light, as well as reflection spectra of meshy structure with (c) linearly polarized incident light and (d) randomly polarized incident light.

close to 0 since about half of the polarization component is not used for the generation of SPR. To suppress the calculation error, more sharp dip curve is desirable. On the other hand, the clear dip of the reflection spectrum is observed in the meshy groove pattern when the randomly polarized light is used as the incident light. The resonant wavelengths in this experiment are listed in Table 4.12. These results verified that randomly polarized component of the incident light can effectively generate SPR by using meshy groove structure. And the polarization filter is not necessary in this case.

Table 4.12 Resonant wavelength and sensitivity of the groove and meshy pattern

	Groove			Cross pattern		
	TM	TE	Random	TM	TE	Random
λ @ n = 1.33 [nm]	623.2	–	615.2	585.8	588.4	588.4
λ @ n = 1.36 [nm]	632.8	–	624.4	594.5	597.0	597.1
Sensitivity [nm/RIU]	**321.6**		**305.7**	**287.9**	**288.3**	**290.0**

4.3.3.3 The effect avoidance of polarization defect

In the nanoimprint SPR biosensor, the incident light goes through the sealing cover of microchannel since the light is irradiated into the front surface of sensor device. Considering the mass production of the sensor device, the sealing cover is desirable to be made of a plastic component or film that sometimes includes the polarization defect. Polarization defect is caused by the one-way alignment of the plastic molecules that have the birefringence. And this defect causes the rotation of polarization of the passing light.

Figure 4.77 shows the experimental setup to analyze the polarization defect of sample device. From the bottom side, a randomly polarized light from the halogen lamp (MHF-FB150LR, Moritex, Japan) is irradiated to the first polarization filter (PH-30, Sigma Koki, Japan). And then, this light goes through the second polarization filter (PH-50, Sigma Koki, Japan). The polarization filtering directions of both polarization filters are set to be perpendicular to each other. The passing light from the first polarization filter is linearly polarized and does not go through the second polarization filter. In this case, no bright light can be observed from the upper side as shown in Fig. 4.77(b). The pictures of the observation results when PDMS plate and acrylic plate are inserted between the first and second polarization filter are shown in Fig. 4.77(c) and Fig. 4.77(d), respectively. A bright light is observed by inserting the acrylic plate, though no bright light is observed by inserting PDMS. This means that the acrylic plate includes the polarization defect. If there is a polarization defect in the sealing cover of sensor device, it is impossible to remove it only by adding the polarization filter in the optical system.

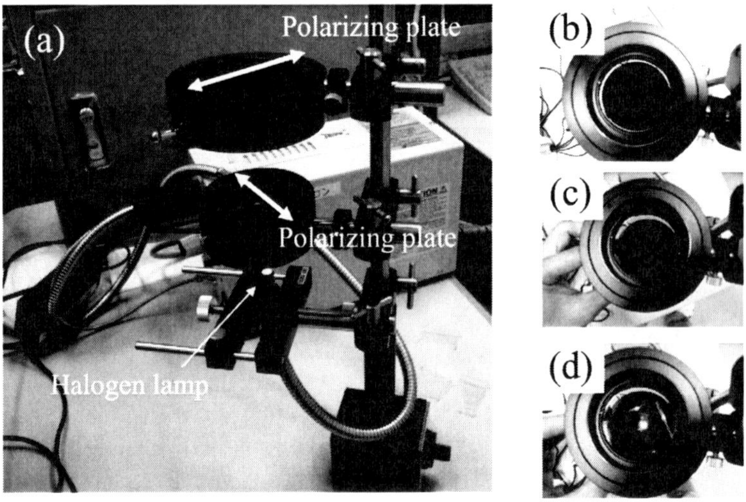

Figure 4.77 Pictures of polarization error evaluation. (a) Experimental setup, (b) no sample, (c) PDMS plate, and (d) acrylic plate between two polarizers that are across each other.

To simulate the sensor device with a polarization defect in its sealing cover, the optical evaluation was carried out by inserting the acrylic plate on the sensor device. Figure 4.78 shows the reflection spectra when TM polarized light and randomly polarized light is inserted as an incident light on the one-dimensional groove and the meshy groove structures. They are evaluated with the sample materials of water and ethanol. It can be observed that the resonant wavelength does not shift very much on the one-dimensional groove structure even when the sample material changes. This means that the sensor with the one-dimensional groove structure cannot work properly when the polarization defect is included in it. In the case of the meshy groove structure with TM polarized light, the reflection curve is slightly deformed since the detection of the reflection light is affected by the spectrometer whose sensitivity has the polarization dependence.

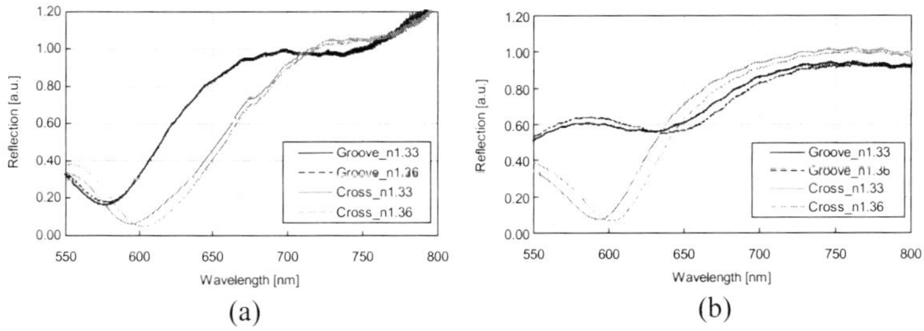

Figure 4.78 Reflection spectra of groove and meshy (cross) structure through the cover plate with the polarization defect with (a) TM polarized and (b) randomly polarized incident light.

When the randomly polarized light is used as the incident light, the resonant curve of the line groove structure is difficult to observe. On the other hand, the normal reflection curve can be observed on the meshy groove structure since the polarization dependence of the spectrometer is not affected by the random polarization light. Table 4.13 shows the resonant wavelength of these results. It is verified that the meshy groove structure with the randomly polarized incident light can work as a usual sensor even in the case that the polarization defect is included within the sealing cover of the sensor device.

Table 4.13 Resonant wavelength and sensitivity of the groove and meshy pattern with the polarization defected cover plate

	Groove		Cross pattern	
	TM	Random	TM	Random
λ @ n = 1.33 [nm]	576.4	–	595.2	592.4
λ @ n = 1.36 [nm]	577.6	–	603.4	601.2
Sensitivity [nm/RIU]	**38.1**	**-**	**274.8**	**294.4**

4.3.4 Pre-Mass Production Test

To put the sensor device to practical use as a commercialized product, the device needs to be fabricated with low cost and high reproducibility. Especially for the biosensor, many error factors such as the probe immobilization error (density, orientation, activity, etc.) and the reaction errors can be included in the consequent processes. Therefore, the fabrication error of the sensor device should be as small as possible. In the conventional localized SPR device using the gold colloids, the fabrication process is easily affected by the chemical stability of the surface of the glass substrate and gold colloids. And the material cost of gold colloid is also a challenging issue. In our nanoimprint SPR sensor, the nanopatterns can be produced by the replication process with high throughput. Since the fabrication processes are completed by the processes established as the mass production techniques, the fabrication error is expected to be small enough to be used for a commercialized product.

We have fabricated a specific number of devices and evaluated their optical characteristics to analyze the fabrication error. Figure 4.79 shows the fabricated sensor chip with the microchannel. Five hundred devices were prepared by injection molding process. TOPAS (5013L-10) was used as an injection polymer. The 1st, 200th, 300th, and 400th devices are extracted and sputtered with 80 nm gold layer. After sealing the microchannel with the pressure sensitive sheet, the optical property was evaluated by filling the microchannel with purified water. The evaluation was conducted at six arrays for each device. Twenty-four arrays in total are evaluated and this result is shown in Fig. 4.80. The average, maximum, and minimum resonant wavelengths are 599.3 nm, 603.9 nm, and 589.0 nm, respectively. Three standard deviations were 9.07 nm, which is about 1.5%. This result is much better than that of the conventional gold colloidal method since the resonant wavelength of the conventional method is directly affected by the sizes of the colloids. It is verified that the fabrication error can be suppressed by using the nanoimprint process from this result. Furthermore, fabrication errors have the potential to be improved by optimizing the parameters of the injection molding.

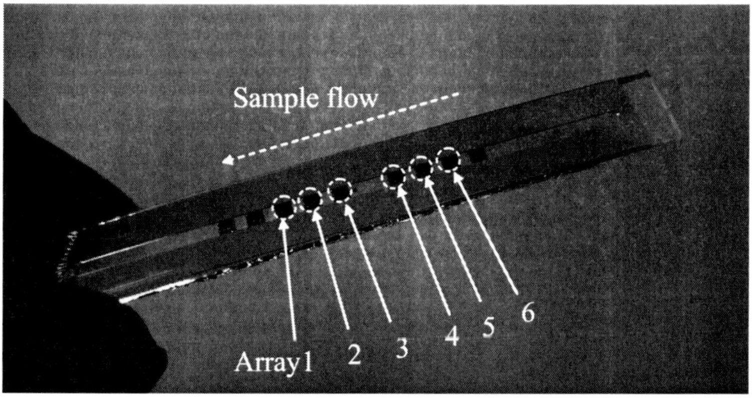

Figure 4.79 Picture of sensor chip fabricated by injection molding for pre-mass production evaluation.

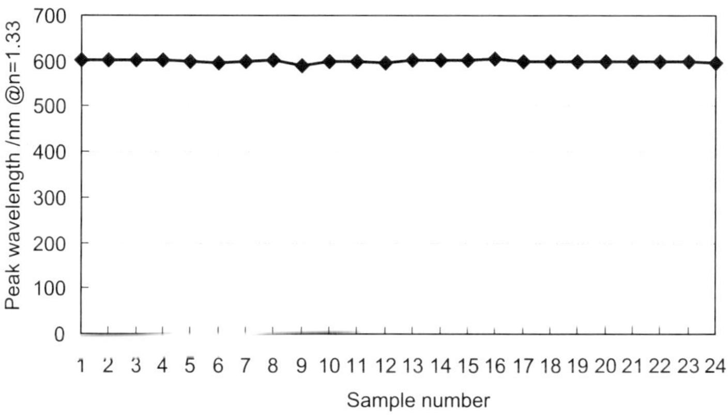

Figure 4.80 Fabrication error evaluation result. The resonant wavelengths of 24 arrays extracted from 500 device samples are measured.

Chapter 5

Proto-Model Systems for Nanoimprint Biosensor

As the equipment to analyze biomolecular interactions, biosensor systems based on surface plasmon resonance (SPR) have already been commercialized. Most of them are, however, very expensive and large to compensate the system's environmental parameters rigorously. On the other hand, though some low-cost SPR systems are also in the market recently, they can be used in restrictive cases since they are generally susceptible to background noise factors. For instance, the refractive index of water changes about 10^{-4} when its temperature varies at 1°C, which is sufficiently large in detecting the biomolecular interactions.

In this study, we have developed a localized SPR sensor based on nanoimprint technology. By using this method, the background noises can be considerably eliminated as verified in the previous chapter. This enables the biosensor system to omit the temperature control system. And the optical system can also be very simple since the total internal reflection of incident light is not necessary in our sensor device. By taking full advantage of these properties, we have developed three different types of proto-model systems referred to as "nanoimprint biosensors". The first one is for laboratory use, the second one is a portable system, and the third one is handy-type system. The details of these systems are presented in this chapter.

Nanoimprint Biosensors: The Fusion of Nanofabrication, Nanophotonics, and Nanobiology
Takeo Nishikawa and Satoshi Fujita

ISBN 978-981-4316-67-5 (Hardcover), 978-981-4463-42-3 (eBook)
www.panstanford.com

5.1 First Proto-Model for Laboratory Use

As the first proto-model, we have developed a system supposed to be used in a laboratory. In the laboratory, the system will be used for analyzing the biomolecular interactions. Considering its usability, the flow control system and the sample injector should be included in the system. The all-in-one system is constructed as shown in Fig. 5.1. The basic specifications are depicted in Table 5.1. The dimension of the system is W250 mm × D250 mm × H150 mm, which is about 10 times smaller than the commercialized conventional SPR system (Biacore-X, GE Healthcare). The features of this system will be presented below.

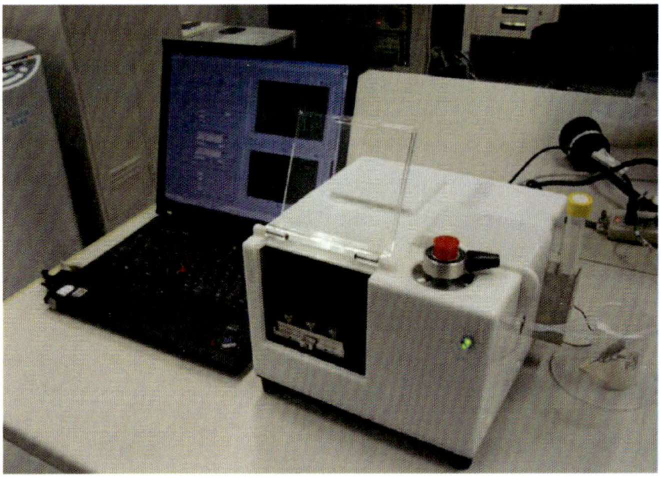

Figure 5.1 Picture of first proto-model of nanoimprint biosensor system.

5.1.1 Optical System

In nanoimprint SPR sensor, the white light is irradiated from the front surface of the sensor device and the reflection spectrum is detected. And the resonant wavelength is calculated. Figure 5.2 shows the optical system, which is constructed on the upper part inside the first proto-model. As a light source, halogen lamp (LS1-LL, Ocean Optics Inc., USA) is used. The light is collimated after passing through an aperture and goes through detachable polarizer and beam splitter (BS CUBE NON-POL VIS 47121, Edmund Optics

Inc., USA). And then, the direction of light is changed to down side by a mirror and is focused by an objective lens (Plan N ×10, Olympus Co., Japan). The practical diameter of a focused spot is several hundred micrometers. The focused light is irradiated on a sensor surface and the reflected light is collected by the objective lens. Subsequently, the reflected light goes upside, is split by the beam splitter, and reaches the spectrometer (USB4000, Ocean Optics Inc.). The reflection spectrum is analyzed by a personal computer connected to the spectrometer.

Table 5.1 Specification of first proto-model

Parameter		Specification
Overview	Principle	Nano-SPR
	Detection limit	Several tens ng/mL
	Array	Max 100 array
	Dimensions	W250 mm × D250 mm × T150 mm
	Weight	About 10 kg
	Control	PC control
	Output	USB
Optical system	Light source	White light
	Detector	Spectrometer
Flow control system	Pump	Electroosmotic pump
	Sample injector	Manual
Sensor chip	Dimension	(48 × 28 × 1.0 mm)
	Surface	Gold
Software	Program	Lab VIEW
	OS	Window XP
	Memory	512 MB

To analyze multiple detection spots or spatial distribution of the reaction, arbitrary multiple points can be detected with the motorized linear XY stage (SGSP15-10, Sigma Koki Co., Ltd., Japan), which is equipped under the sensor holder. To cool off the interior of the system especially taking into consideration the heat from the halogen lamp, two cooling fans (MDS410-12, Oriental Motor Co., Ltd., Japan) are equipped to evacuate the heat.

Figure 5.2 Picture of the optical system inside first proto-model.

5.1.2 Flow Control System

In the biosensor system, a small amount of target molecule needs to be measured with very high sensitivity. The flow control system which controls the sample solution is, therefore, a very important component in the system. Above all, the flow pump is a key device in the flow system. The demands for the flow pump in our biosensor system can be summarized as below.

- Small size
- No pulsating flow
- Flow rate can be adjustable in the range of 20–200 μL/min
- Adaptive pressure loss is from a few to 100 kPa
- No air bubble generation

To realize the flow control system which satisfies the above demands, we have compared the commercially available flow pumps as the first step. Figure 5.3 shows an overview of the typical commercialized pumps in terms of the flow rate and adaptive pressure loss. It can be seen that the pumps satisfying the above demands of flow rate and pressure loss are syringe pump, plunger pump, and electroosmotic pump. In this part, single plunger pump and electroosmotic pump are evaluated since the syringe pump is too large and too expensive to integrate in the low-cost biosensor system.

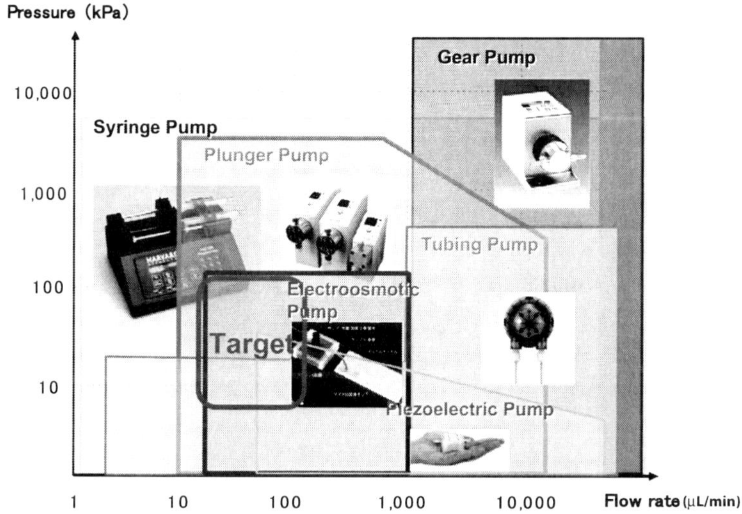

Figure 5.3 Comparison of the typical commercialized pumps in terms of the flow rate and adaptive pressure loss.

5.1.2.1 Single plunger pump

In a plunger pump, a piston so-called as "plunger" moves back and forth with a valve being switched on and off. The sample solution is sucked in and poured out by means of the plunger motion. As the pulsating flow is generated by the back-and-forth motion of plunger, double plunger pump whose two plungers move to counteract the pulsating flow each other is proposed and is used for techniques such as HPLC. However, a single plunger pump with a pulse damper is evaluated in this study since the double plunger pump is expensive and large in size in general.

As a single plunger pump, a small-sized commercially available one (Leafbird-200SZW, Uniflows Co., Ltd., Japan) shown in Fig. 5.4 is selected. As shown in a schematic image of Fig. 5.5, a flow sensor (ASL1430-24, Sensirion, USA) is connected to the pump and the flow rate is monitored by a personal computer. The results with and without the pulse damper (PD0235S, Uniflows) between the pump and the flow sensor are analyzed. As a comparison, the result by a syringe pump (PHD2000, Harvard, USA) is also presented. The flow rate is set at 20 μL/min in all cases. The result is shown in Fig. 5.6. In the syringe pump, the average flow rate (0–300 sec) and 3σ were 19.4 and 0.90 μL/min (about 4.6% to the average flow rate), respectively. In the single plunger pump without a pulse damper, an extreme oscillation occurs at an interval of about 10 sec which corresponds to the plunger's motion. In consequence, its 3σ was as much as 66.5 μL/min (about 336.9% to the average flow rate), though its average flow rate is 19.7 μL/min, which is almost same as the syringe pump. The pulsating flow such as this can cause the vibration of the system or instability of the chemical reaction. By inserting the pulse damper, the average flow rate and 3σ were 18.7 and 2.6 μL/min (about 14.1% to the average flow rate), respectively. It is found that the pulsating fluctuation can be reduced 1/25 times compared to the result without the pulse damper. Although the pulsating flow cannot be suppressed perfectly, relatively small pulsating flow can be achieved by a small-sized and low-cost system by this method. One thing which should be cared about when using a plunger pump with a small plunger volume is that the temperature of the motion system can rise due to quick and continuous motion. Since the temperature variation causes the noise and drift factors in the system, it needs to be controlled by a cooling fan.

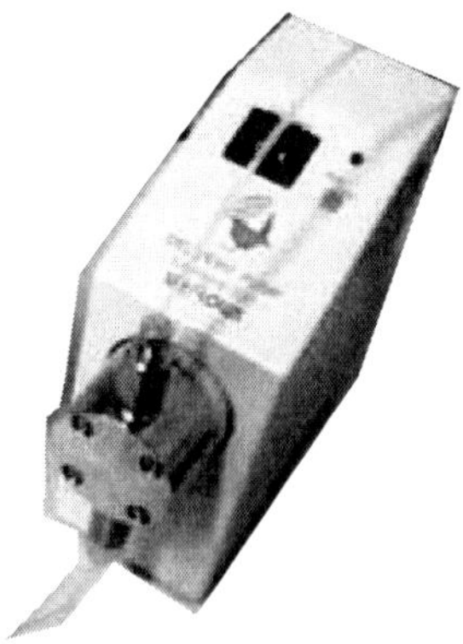

Figure 5.4 Picture of a small single plunger pump used in the experiment.

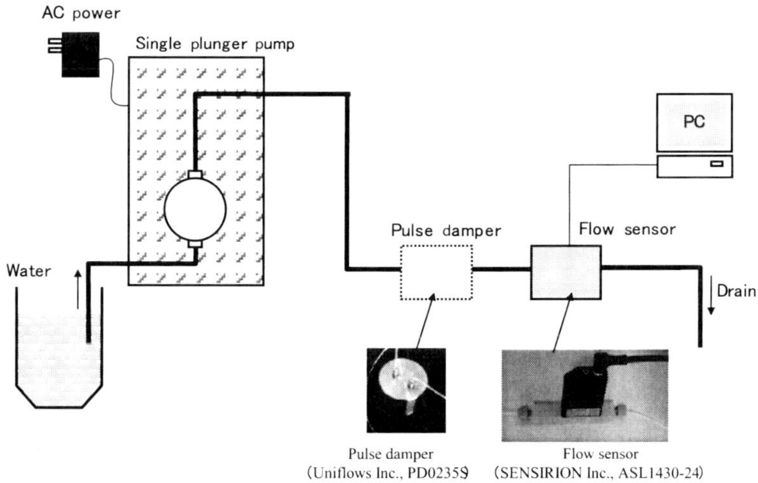

Figure 5.5 Schematic of experimental setup using the single plunger pump.

5.1.2.2 Electroosmotic flow pump

As another candidate, electroosmotic flow (EOF) pump is evaluated in this part. The principle of EOF pump is illustrated in Fig. 5.7. The surface of the dielectric material contacting the liquid sample takes a charge. And the electro-double layer is formed near it. In result, the liquid sample near the dielectric wall also takes a charge. When the electric field is applied on this condition, the charged solution starts to drift. The EOF pump generates no pulsating flow and can send the liquid sample with a high pressure, although its size is as small as fingertip.

Figure 5.8 shows an EOF pump (RP7SP, Nano Fusion Technology, Japan), which has been commercialized. This pump is connected as shown in Fig. 5.9 and the flow rate stability is analyzed. The result at the applied voltage of 3.7 V is shown in Fig. 5.10. In the period from 0 to 300 sec, there is no pulsating flow and a stable flow control is realized. The average flow rate and its 3σ were 23.0 and 1.3 μL/min (5.8% to average flow rate), respectively. Since EOF pump is not a constant volume controlling type, the average flow rate needs to be adjusted by tuning the applying voltage. From this result, it is verified that the no-pulsating flow can be achieved by EOF pump.

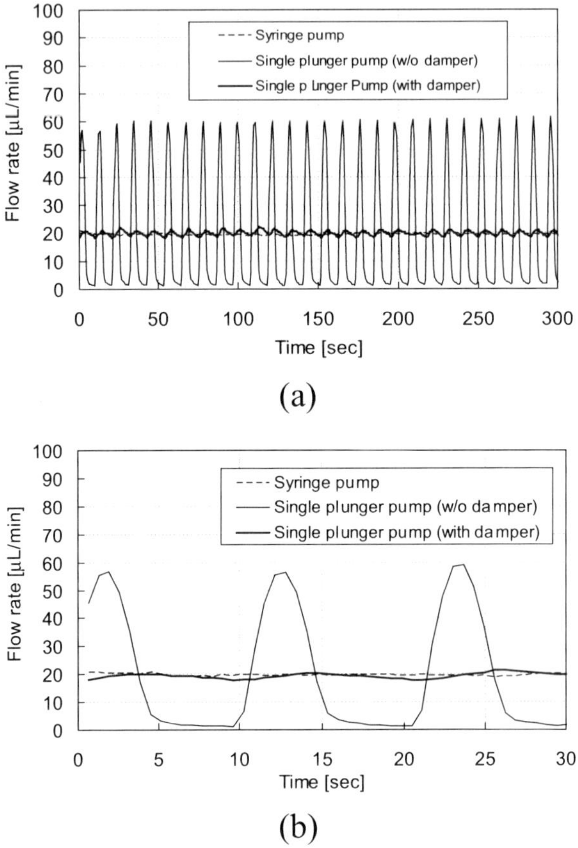

Figure 5.6 Experimental result of the flow rate with syringe pump and single plunger pump in the time range of (a) 300 sec and (b) 30 sec.

However, small fluctuations are observed at about t = 140 and 170 sec. To analyze in more detail, an evaluation in a long period was conducted as shown in Fig. 5.10b. This result shows that some large fluctuations exist in long span. Especially at about 2000 sec, the flow rate dropped down to 0 and recovered at about 2080 sec. This phenomenon was not observed in other pumps such as syringe pump and single plunger pump. In consequence of further studies, it was revealed that the air bubble is generated at the electrodes inside the EOF pump by the electrolytes and that it remains inside the flow sensor for a while. In the practical use in the laboratory, the

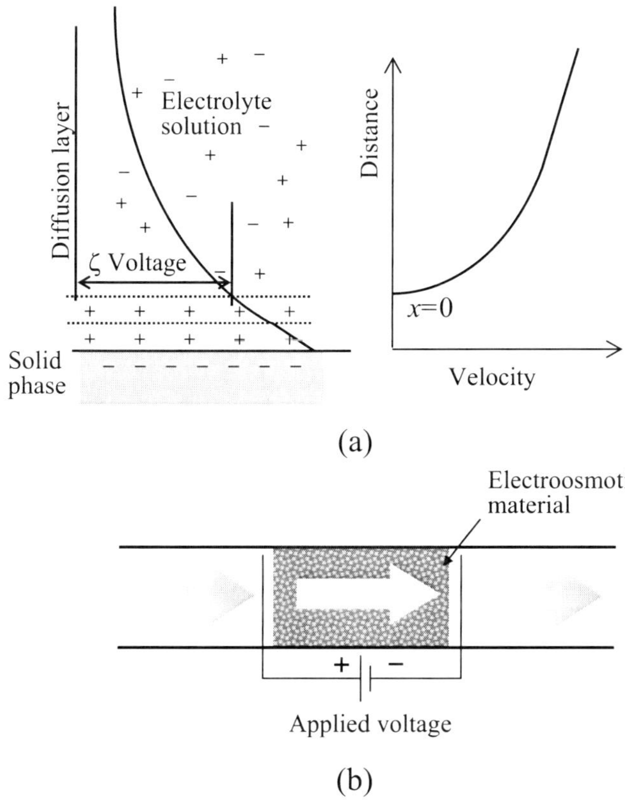

Figure 5.7 Schematics of electroosmotic flow (EOF) principle: (a) cross-sectional image and (b) configuration of EOF pump.

experiment will continue for a few hours and the flow instability due to the air bubble can be a fatal problem.

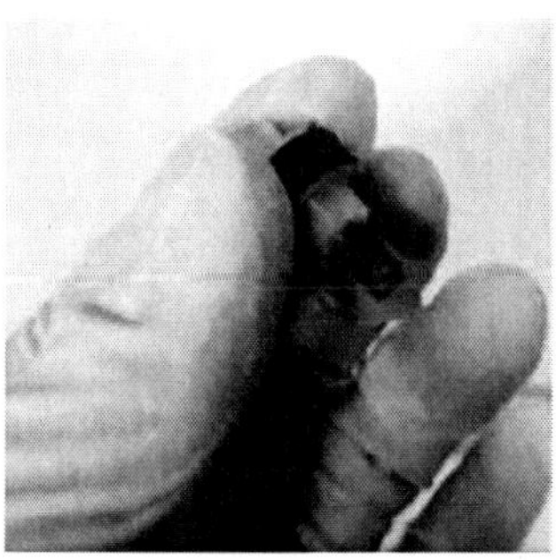

Figure 5.8 Picture of electroosmotic pump.

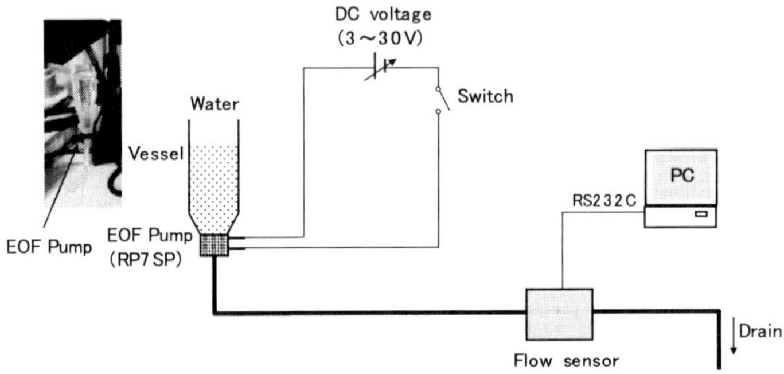

Figure 5.9 Schematic of the experimental setup of the EOF pump.

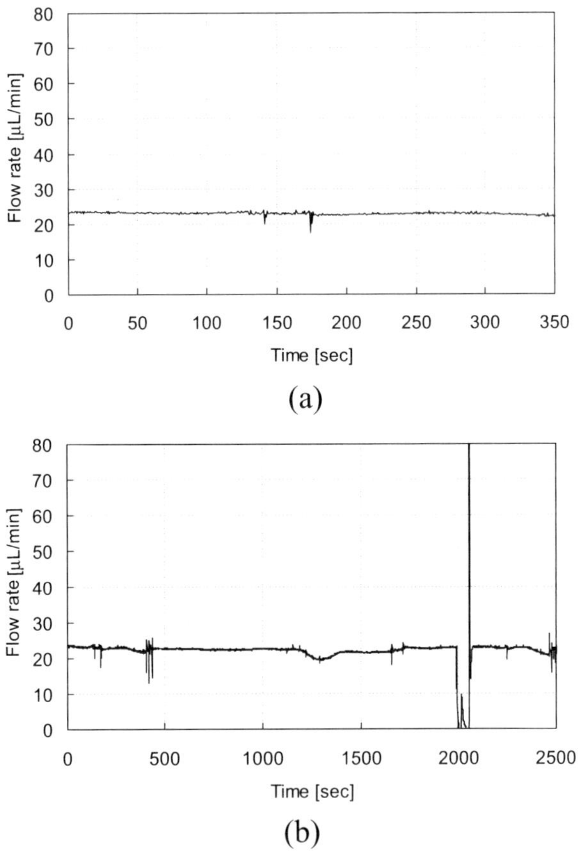

Figure 5.10 Experimental results of EOF pump evaluation in (a) short period and (b) long period.

To avoid the air bubble, a large amount of sample loop is inserted in the flow control system as shown in Fig. 5.11. The solution in the sample loop is indirectly driven by EOF pump. In this method, the sample solution directly driven by the EOF pump does not get to the flow sensor until the solution inside the sample loop is all pushed out. Therefore, the air bubble generated at the EOF pump does not affect the flow rate signal. The solution inside the sample loop can be prepared and reloaded easily by using a sample injector (Manual sample injector 9725i, Rheodyne, USA). The volume of the sample loop should be selected by calculating the flow rate and the requisite period for the experiment. The experimental result under the constant applied voltage of 3.7 V is shown in Fig. 5.12. The average flow rate and its 3σ were 22.8 and 0.48 μL/min (2.1% to the average flow rate), respectively.

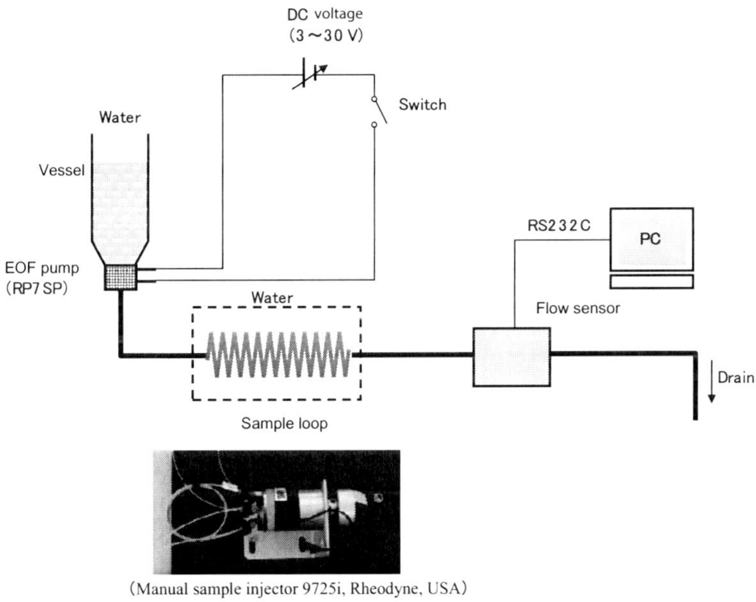

Figure 5.11 Experimental setup of the indirect flow control system using EOF pump.

As shown in this result, by using the EOF pump and indirect driving technique, high stable flow control can be realized. The indirect driving method can also avoid the precipitation of salt at electrodes even if the buffer solution includes the salt. Furthermore, the sample injector can achieve high usability in loading the sample solution between the EOF pump and sensor device.

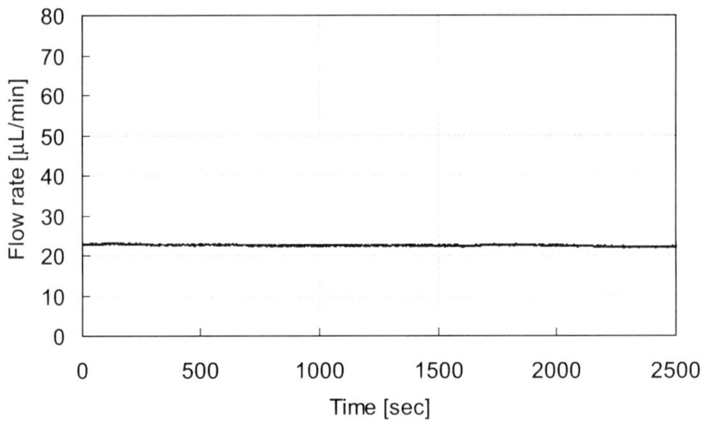

Figure 5.12 Experimental result of the indirect flow control using EOF pump in a long period.

5.1.2.3 Integration to the proto-model

As described above, two types of pumps (single plunger pump and EOF pump) are evaluated. In this proto-model, EOF pump is selected and integrated into the system for its flow stability and compactness. The schematic image of the flow control system in the proto-model is illustrated in Fig. 5.13. EOF pump with the water vessel is set at the most upstream of the flow system. After that, a sample loop for the buffer solution with a manual injector (Manual sample injector 9725i, Rheodyne) is prepared. A flow sensor to monitor the flow rate (ASL1430-24, Sensirion) and a sample injector to insert the sample solution such as proteins exist after that. And a sensor device is set at the downstream. The sensor device can be easily connected to the flow system by using O-rings on the sensor stage. The picture of the actual flow control system is shown in Fig. 5.14. The space for the flow control system was about W80 mm × D150 mm × H150 mm.

The voltage applied on the EOF pump can be set at a constant value. In addition, it can also be controlled automatically to achieve a specific flow rate by a feedback control loop. Figure 5.15 shows the experimental result with setting specific applied voltages. It is observed that the flow rate changes within several

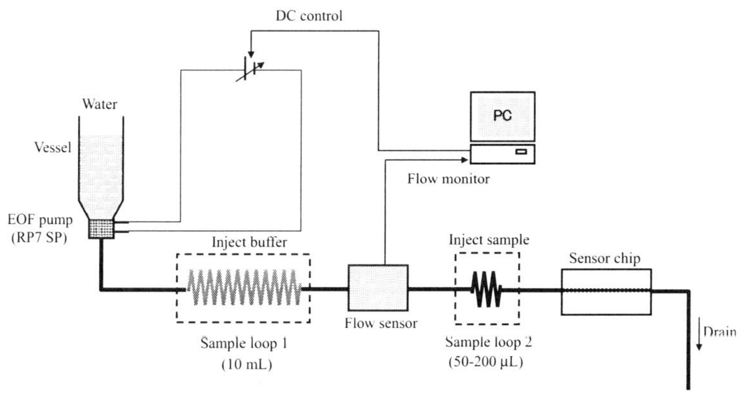

Figure 5.13 Schematic of feedback flow control system adopted inside the first proto-model.

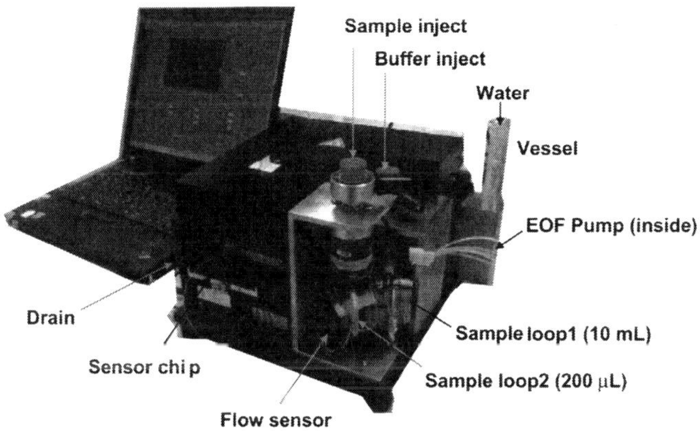

Figure 5.14 Picture of flow control system built in the first proto-model.

seconds according to the applied voltages. The flow rate with the constant applied voltage varies depending on the conditions of individual pump and the liquid solution (temperature, pH, etc.) and the pressure loss of the flow system. Therefore, for a constant and high reproducible flow control, the feedback control of the applied voltage is desirable and useful. To achieve the targeted flow rate, the actual flow rate is compared with the targeted flow rate and the proportional-integral-derivative (PID) control is conducted. The feedback system is constructed by LabVIEW (National Instruments, USA) program. The experimental

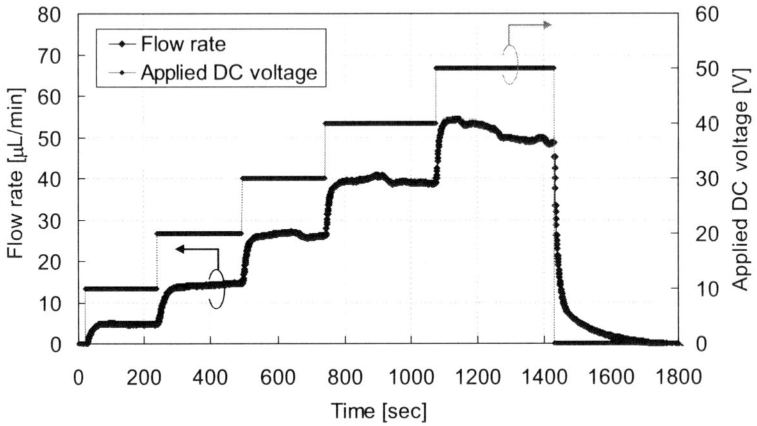

Figure 5.15 Experimental result of flow control system in first proto-model changing the DC voltage applying on the EOF pump.

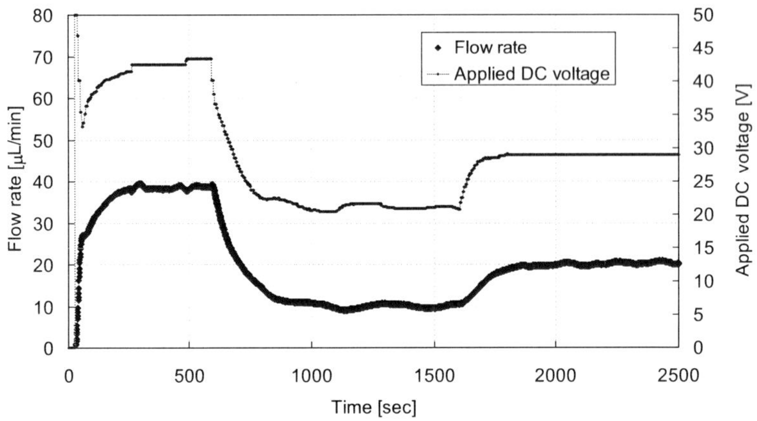

Figure 5.16 Experimental result of flow control system in first proto-model changing the target flow rate. The target flow rate is set at 40 μL/min between 0 and 600 sec, 10 μL/min between 600 and 1600 sec and 20 μL/min after 1600 sec.

result to evaluate the feedback response is shown in Fig. 5.16. The target flow rate was set as 40 μL/min from 0 to 600 sec, 10 μL/min from 600 to 1600 sec, and 20 μL/min after 1600 sec. When the target flow rate was changed, the actual flow rate changed smoothly and was stabilized at the target flow rate. It was observed that there was no overshoot in these results though it takes about 250 sec to achieve the target flow rate. The feedback experimental result to evaluate the stability in a long time period (about 12,000 sec) is shown in Fig. 5.17. The target flow rate was set at 20 μL/min. It was observed that the applied voltage is automatically adjusted to keep the flow rate constant. In result, the average flow rate and 3σ between 250 and 12,000 sec are 19.9 and 1.6 μL/min (7.9% to the average flow rate). This result confirms that a stable flow control can be achieved by using an indirect EOF pump with a feedback control. As proved in these experiments, we have constructed a small-sized, high stable, and accurate flow control system for a biosensor system based on EOF pump.

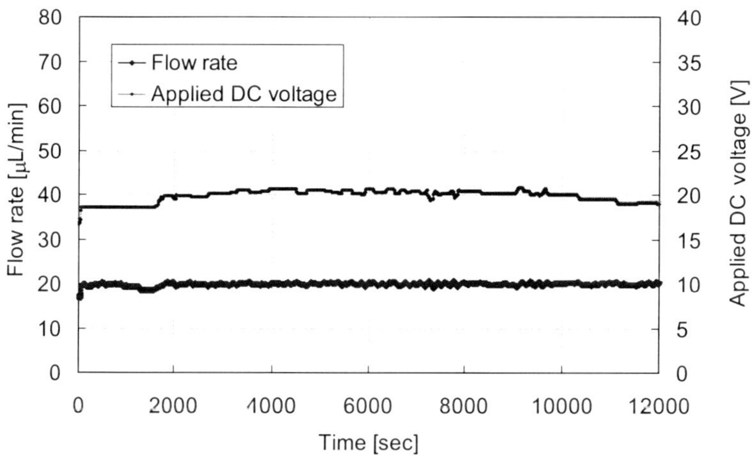

Figure 5.17 Experimental result of flow control system to evaluate the flow rate stability in a long time period.

5.1.3 System Control Software

This proto-model can be controlled by a computer program constructed based on LabVIEW ver7.0 (National Instruments). The display image is shown in Fig. 5.18. The main functions of the program are mentioned below.

1. Calculation and storage of the resonant wavelength data
2. Control of motorized linear XY stage
3. Control of flow control system

4. User interface (input of the parameters and display of output data)

The detail of each function is introduced hereafter.

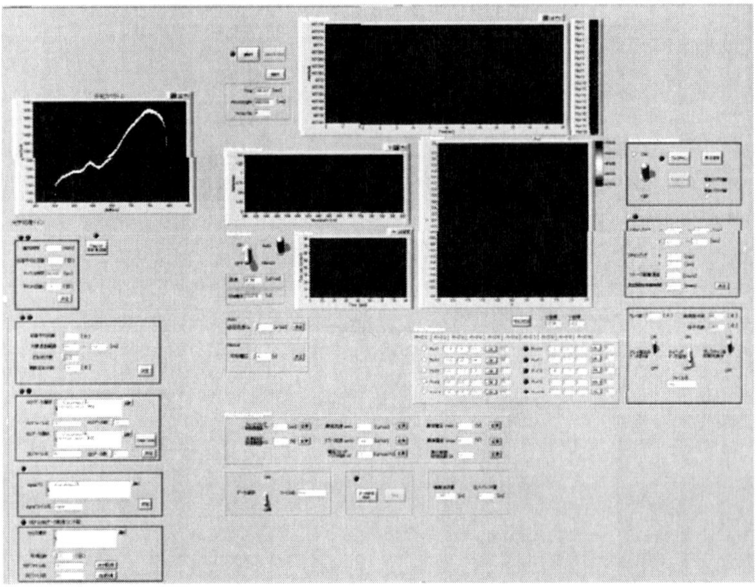

Figure 5.18 Desktop image of the control software for the first proto-model.

5.1.3.1 Calculation and storage of the resonant wavelength data

To use a biosensor, the resonant wavelength of the sensor chip needs to be calculated with high accuracy. In this system, the resonant wavelength is calculated from the reflection spectrum data acquired by a spectrometer (USB4000, Ocean Optics). The calculation step is introduced in the flow chart shown in Fig. 5.19. In the first instance, the setup parameters such as the exposure time of the spectrometer and data acquisition number for average calculation are selected. To calculate the reflection spectrum $R_t(\lambda_i)$, the reference spectrum $\mathrm{Rf}(\lambda_i)$ and background spectrum $\mathrm{BG}(\lambda_i)$ are necessary. They can be previously prepared by getting the spectrum data with putting a mirror and absorbance material (black paper, etc.), respectively. Then the reflection spectrum is calculated by the following equation.

$$R_t(\lambda_i) = \{\text{Sig}_t(\lambda_i) - \text{BG}(\lambda_i)\} / \{\text{Rf}(\lambda_i) - \text{BG}(\lambda_i)\}$$

where $\text{Sig}_t(\lambda_i)$ is the signal spectrometer with putting the sensor chip. To find the resonant wavelength with a higher resolution than the resolution of the spectrometer, the local minimum wavelength is calculated by using a polynominal approximation method. In the polynominal approximation, the calculation parameters such as the order of polynominal function and the wavelength range for the approximation are significant to achieve high accuracy and low noise. The calculation is repeated with the preset interval time. The calculated data can be stored in *ACCII* format. When the stop button is pressed, this calculation flow is stopped.

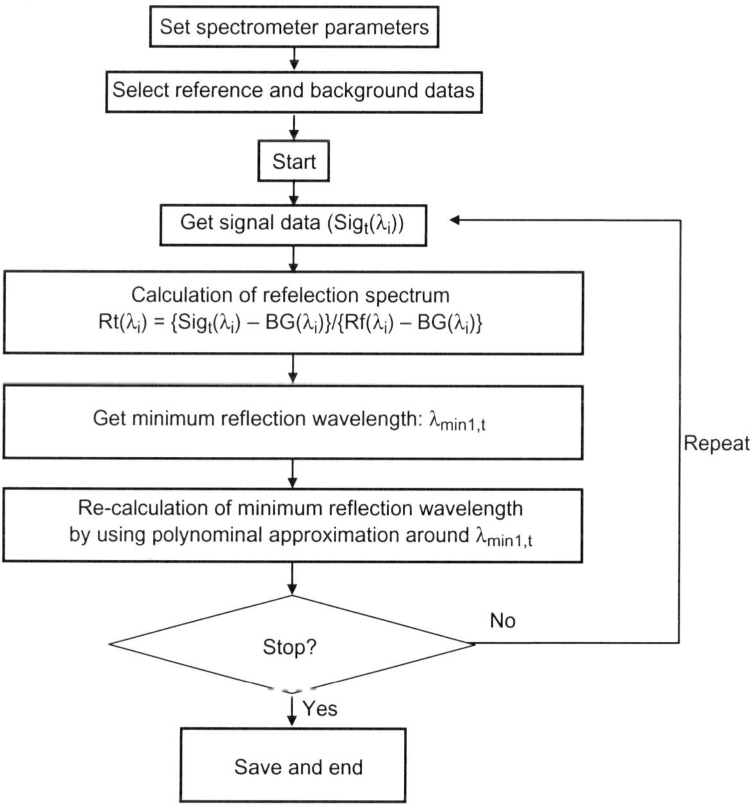

Figure 5.19 Flow chart for calculating the peak resonant wavelength in the first proto-model.

5.1.3.2 Control of the motorized linear XY stage

To control the sensing point, motorized linear XY stage (SGSP15-10, Sigma Koki Co., Ltd., Japan) is equipped under the sensor device holder. The arbitrary points not greater than 100 for calculating the resonant wavelength can be chosen by clicking the positions on the map in the display or by inputting the X–Y coordinate and each diameter. It needs to be mentioned that the larger the interval time gets, the more the position number is selected. By using this function, the multiple array data can be acquired. Furthermore, the reaction distribution on the sensor surface area can also be detected.

5.1.3.3 Flow control system

The flow control setting can be done by this program. There are two control modes. One is the manual mode in which the constant voltage is applied on the EOF pump. And the other is the auto mode in which the applied voltage is controlled to keep a constant flow rate. The parameters for the PID feedback control can also be set in this program. As shown in this subsection, we have constructed the system control program based on LabVIEW. This can provide useful and easy operation for system users. The examples demonstrated by this system will be introduced in the following chapter.

5.2 Second Proto-Model for Portable Use

As the applications of the biosensor, the medical diagnosis, the food analysis and the environmental monitoring, and so on are expected. If the system is portable, it can be used in a wide variety of locations. As the second proto-model, a portable and easy operational system is constructed. In this system, the rapid tests of a specific binding reaction are targeted while the various experiments are targeted in the first proto-model. The picture of this system is shown in Fig. 5.20. Figure 5.21 shows its inside. The system size is scaled down to W160 mm × D160 mm × H144 mm since the optical system is optimized and one-directional motorized linear stage is adopted. Its weight is about 3 kg, which is five times lighter than the first proto-model. The downsizing and weight saving make it possible to use it on sites.

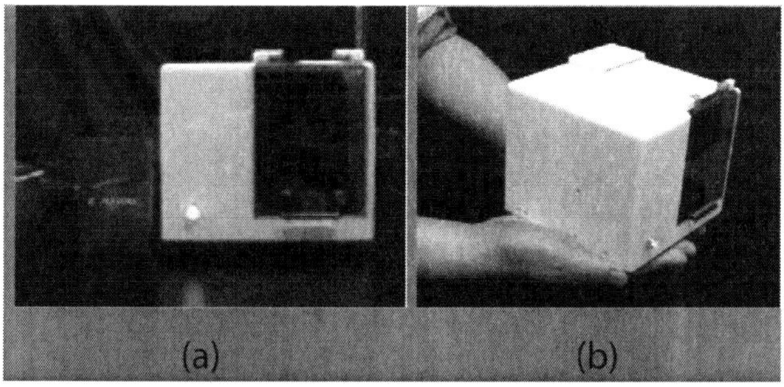

Figure 5.20 Pictures of second proto-model (a) compared with a cigarette box and (b) on hand.

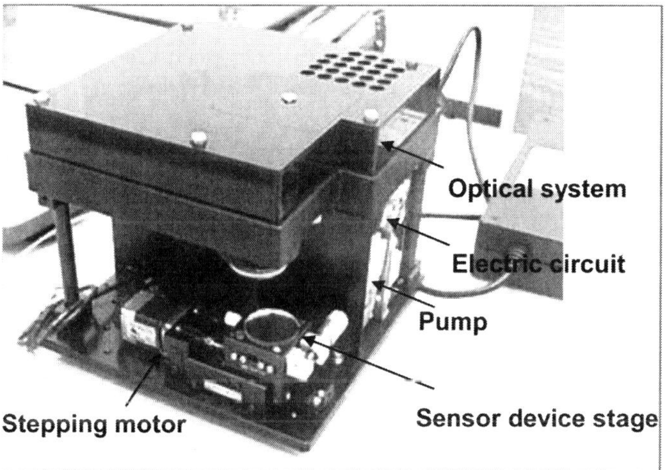

Figure 5.21 Pictures of second proto-model removing the outer case.

5.2.1 Optical System

The optical system is shown in Fig. 5.22. Although its basic mechanism is same as the first model, it realizes downsizing by optimizing its layout. As the light source, a halogen lamp (LS1-LL, Ocean Optics) which is in the first model is used with removing a steel case and adding an original cooling wheel. A spectrometer (USB4000, Ocean Optics) is arranged in close proximity of the half prism to cut down the extra space.

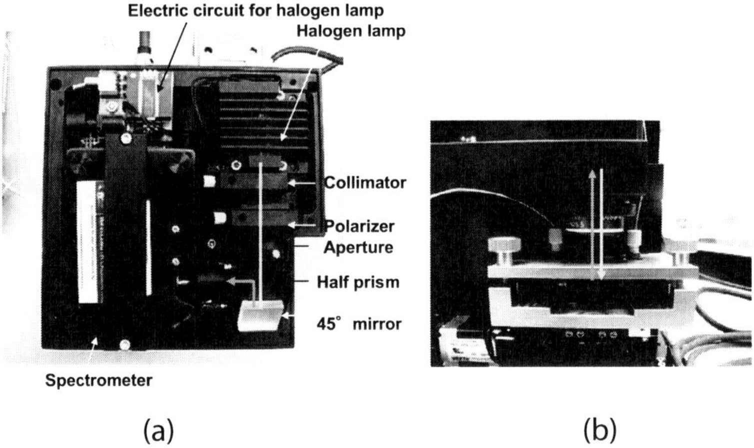

Figure 5.22 Pictures of optical system inside the second proto-model: (a) top view and (b) magnified side view.

5.2.2 Flow Control System

The concept of the flow control system in the second proto-model is changed from that of the first proto-model. The schematic image of the flow control system is shown in Fig. 5.23. In the second proto-model, a target sample solution is directly injected in the sensor device and it is sucked by a compact suction pump (PP100PZ23, Denso Sangyo, Japan). In this system, the polluted sample solution such as blood and urine does not enter the flow control system if a buffer tank is prepared after the sensing point on the disposable sensor device. Therefore, the cleansing of the flow path is much easier than the first mode. The direction of the solution flow is controlled by two electromagnetic valves (LFVA1210120H, The Lee Company, USA) which is prepared at both sides of the suction pump. The picture of the flow system is shown in Fig. 5.24.

5.3 Third Proto-Model for Handy Use

To be used in the wider fields, a simpler, smaller, and more inexpensive system is desirable. As the third proto-model, we have constructed a palm-sized biosensor (Fig. 5.25).

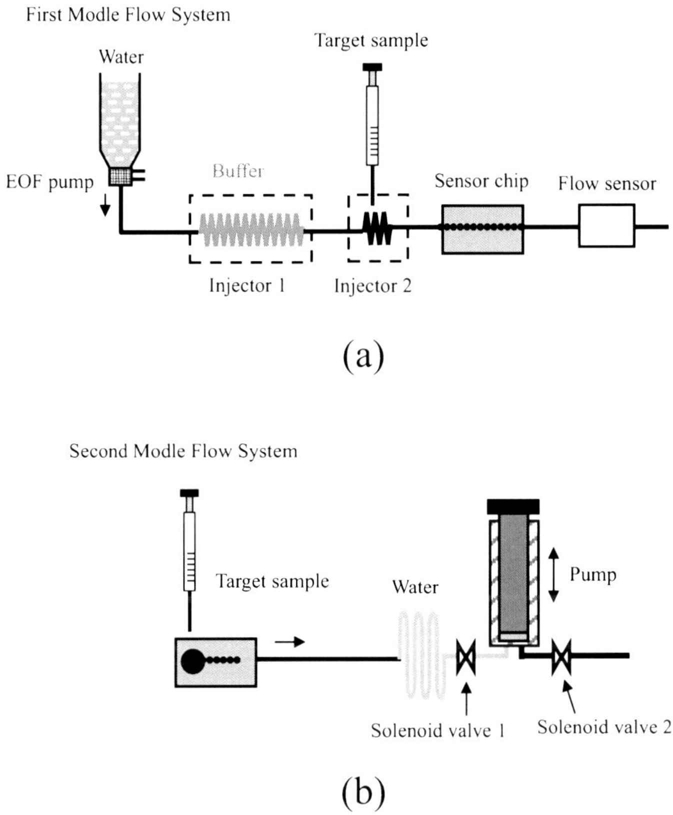

Figure 5.23 Schematics of flow control systems in (a) first proto-model and (b) second proto-model.

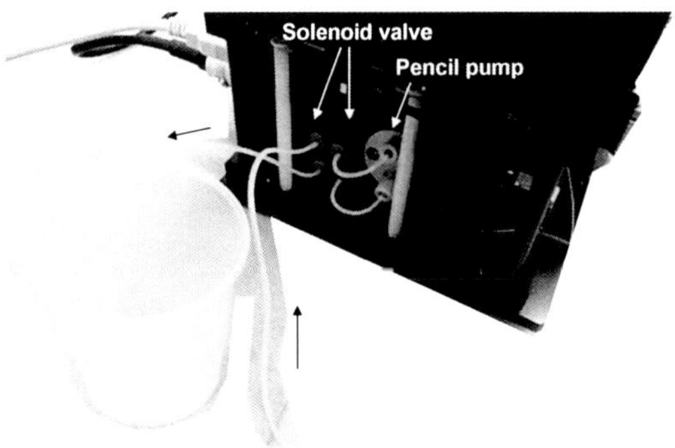

Figure 5.24 Picture of flow control system in second proto-model.

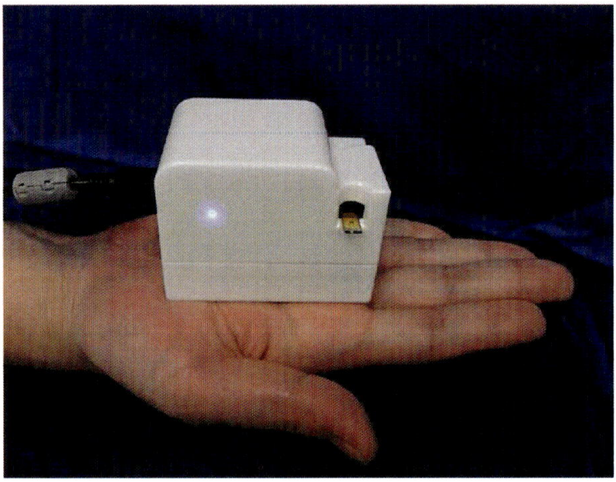

Figure 5.25 Picture of third proto-model.

The features of this system are given below.

- Small size and lightweight (W77 mm × D52 mm × H56 mm, 240 g).
- Four arrays can be detected at once.
- Rapid and easy use.

The details of this system will be introduced in this subsection.

5.3.1 Optical System

In the first and second proto-models, a white light from the halogen lamp is used as a light source and the resonant wavelength is detected by the reflection spectrum. Though this configuration provides a low noise calculation by using a polynomial approximation for the spectrum data, it is difficult to downsize it to a palm size due to the occupied dimensions by the halogen lamp and the spectrometer. In the third proto-model, a diode laser with a specific wavelength is used as a light source and the reflection intensity is detected to observe the reaction on the sensor surface. This configuration can drastically downsize the system and reduce its price. Since this system is subject to the noise and drift, its optical system and control and calculation program need to be designed with taking those factors into consideration.

5.3.1.1 Selection of light source

For realization of a palm-sized biosensor, the dimension and heat emission of the light source need to be as small as possible. And its intensity should be large enough to discriminate the background noise. From these points of view, we have selected a laser diode as a light source for the third proto-model. In the previous models, the reaction of the biomolecules is detected by the resonant wavelength shift. When using a laser diode which has a very narrow emission wavelength band, the reaction of the biomolecules will be detected by the intensity change of the reflected light. Essentially, the resonant wavelength shifts to a longer wavelength by the biomolecular binding reaction. This means that the reflection intensity in the longer wavelength side from the resonant wavelength decreases when the reaction occurs on the sensor surface. On the other hand, it increases in the shorter wavelength side. However, if the wavelength shift is too large, the reflection intensity starts to increase in the longer wavelength side, which means that the quantitative analysis gets complex. Therefore, the light source with a wavelength peak in the shorter wavelength side from the resonant wavelength should be selected. In this development, the light source whose peak wavelength is 635–650 nm is selected. And the resonant wavelength of the nanoimprint biosensor is tuned up to be 650–700 nm by adjusting its structural parameters.

To achieve a highly sensitive sensing, the stability of the light source is one of the most important factors in selecting the device. In this development, we have compared the characteristics of five commercially available laser diodes (Table 5.2). The evaluation is carried out by measuring the power intensity (1830-C, Newport) of the diodes and by analyzing their stability. Figure 5.26 shows the result. The laser diodes are turned on at $t = 0$ and the 3σ from $t = 100$ to 1800 sec is calculated. This result is shown in Fig. 5.27. This result verified that the laser diode of model (b) is most stable. Though other lot products of the model (b) are also evaluated, they had a good stability as this result. These results proved that the stability of the laser diode is very different from each product model. And interestingly, a cheaper model had a higher stability than much more expensive model in this case. With the additional experiments, the stability can be improved by wrapping the light source with silicon and polymer films (Fig. 5.28) to prevent the effect of the temperature fluctuation. The experimental result with each three times average is shown in Fig. 5.29. It is observed that the stability is much improved by wrapping the light source. This wrapped light source is integrated in the system in this development.

Table 5.2 List of laser diodes evaluated in this experiment

Model	Maker	Size	Wave-length	Output power	Voltage	Price
a	A. Co.	ϕ12 mm–L27 mm	650 nm	Less than 1 mW	DC 3 V–40 mA	US$ 5
b	K. Co.	ϕ10.55 mm–L27.8 mm	650 nm	Less than 1 mW	DC 3 V–50 mA	US$ 6
c	G. Co.	W14 × H14 × L24 mm	635 nm	0.8 mW	DC 3–6 V/40 mA	US$ 100
d	G. Co.	ϕ7 mm–L24 mm	635 nm	1 mW	DC 3–6 V/30–90 mA	US$ 90
e	G. Co.	ϕ11 mm–L37 mm	635 nm	1 mW	DC 3–6 V/30–90 mA	US$ 120

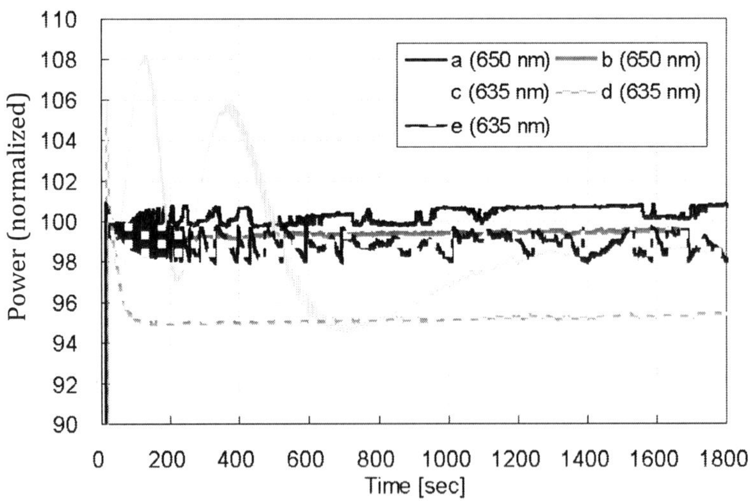

Figure 5.26 Experimental result of evaluating the stability of laser diodes.

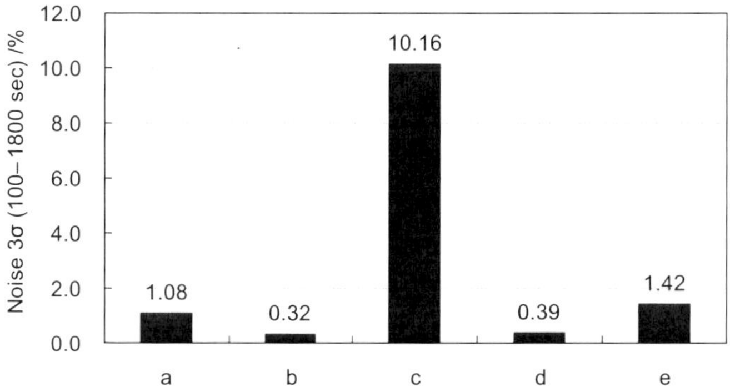

Figure 5.27 Summary of stability evaluation of laser diodes.

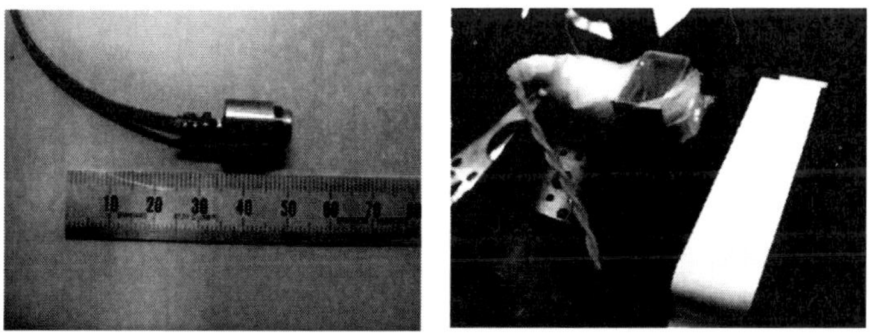

(a) (b)

Figure 5.28 Pictures of laser diode (a) initial condition and (b) after being wrapped by silicon and polymer film.

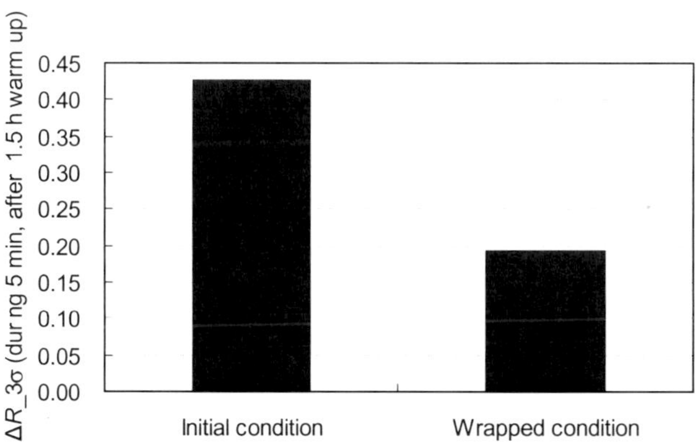

Figure 5.29 Comparison of the continuous stability before and after wrapping.

5.3.1.2 Selection of detector

A detector is also an important device to dominate the signal-to-noise ratio of the biosensor. To achieve miniaturization and high sensitivity (low noise), a photodiode in which a signal amplifier is integrated is desirable. Since the wavelength of the light source is about 650 nm, the detection wavelength range needs to be in a visible light range. We have selected a photodiode (S8745-01, Hamamatsu Photonics Co.) meeting the above conditions (Fig. 5.30). The characteristics of the output voltage to the light intensity are shown in Fig. 5.31. A high linearity is observed in the range of 10–70 μW in the case that the external resistor is 1.0 MΩ. Since the intensity of the incoming light to the detector is actually decided by the system factors such as light source, optical path, reflection coefficient of mirror, etc., the external resistor is arranged to be easily changed to tune up the sensitivity of photodiode.

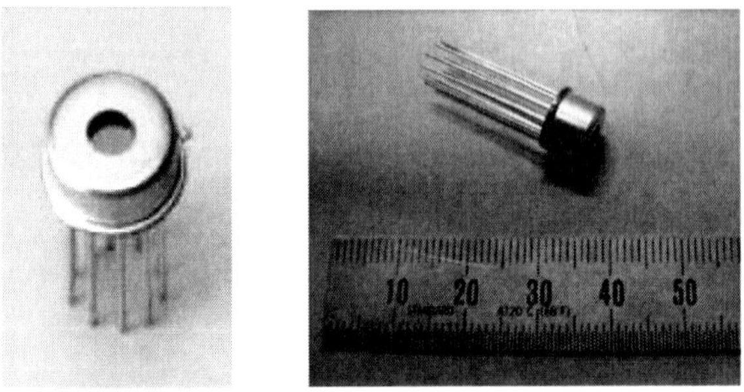

(a) (b)

Figure 5.30 Picture of selected photodiode as the detection device in third proto-model.

5.3.1.3 Optical system

The basic optical path of this system is designed and simulated by using simulation software based on a ray tracing method (Opticad 9, Opticad Co., NM, USA). The result is shown in Fig. 5.32. The light from the source is expanded by a concave cylindrical lens whose focal length is 15 mm. And then, it is collimated by a convex cylindrical lens again after the reflection on the mirror. In result, the original beam width is expanded about 2.7 times (= 40/15) larger for irradiating four array spots. The collimated light is split by a half prism and then reflected upward by a mirror. Subsequently, it enters into four focusing lenses (f = 8 mm, D = 2.0 mm) after an aperture with four holes. Four beams are irradiated on each sensing spot on the sensor chip and then go back to the half prism on the same path. The beams through the half prism reach the photodiodes and their intensity is detected. The picture of the actual optical system constructed on the third proto-model is shown in Fig. 5.33 though the laser diode is not wrapped by film yet in this case.

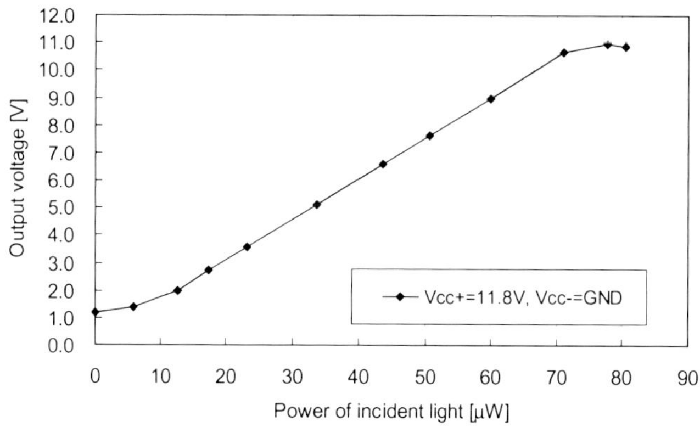

Figure 5.31 Output characteristic of the photodiode as a function of incident light power.

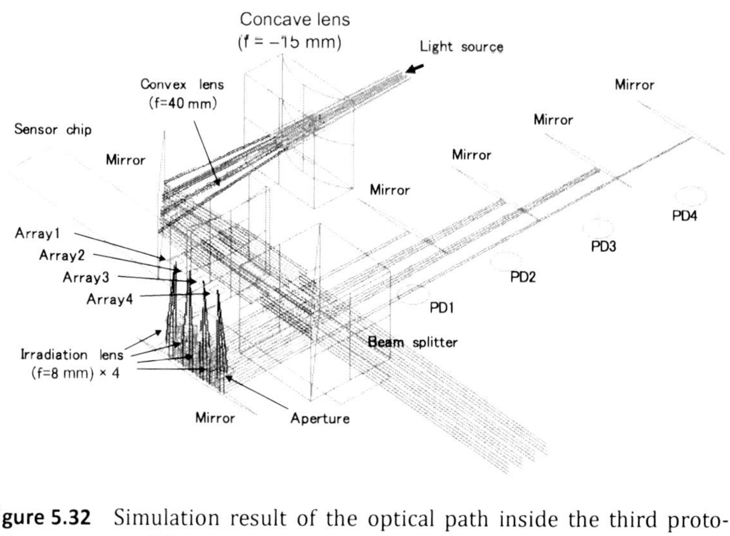

Figure 5.32 Simulation result of the optical path inside the third proto-model.

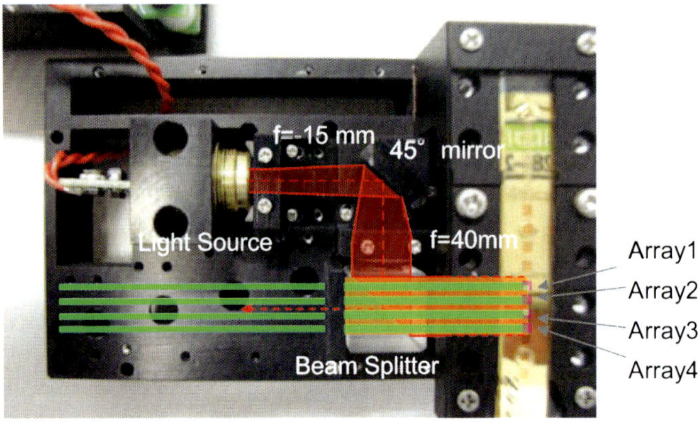

(a)

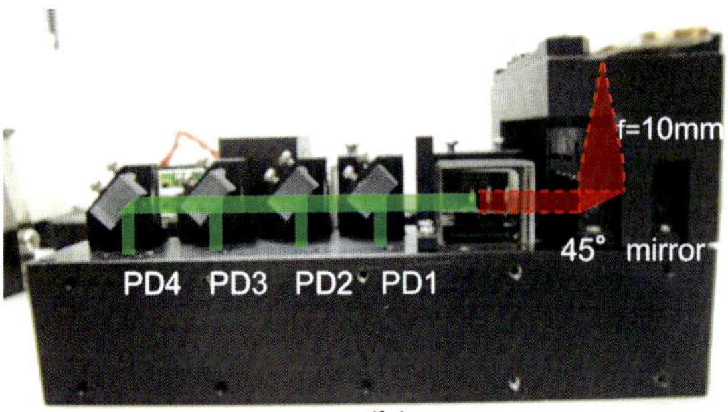

(b)

Figure 5.33 Pictures of optical system in the third proto-model (a) top view and (b) side view.

5.3.2 Microflow System

To realize a rapid and easy test with a palm-sized biosensor, the sample solution is directly injected into the sensor device by using an injection syringe as shown in Fig. 5.34. The injected solution flows in the microchannel on the device and is exhausted from the outlet (Fig. 5.35). The device is fabricated by an injection molding method and pressure sensitive film as described above. By cutting out the pump and flow system in the system, an extensive miniaturization can be realized. Furthermore, the system is not polluted by the sample solution. Accordingly, the onerous cleaning is not necessary in this system just by using a disposable sensor device. As an option, an extra pump can be attached and long-term data can be acquired if a sequential experiment is demanded in this system.

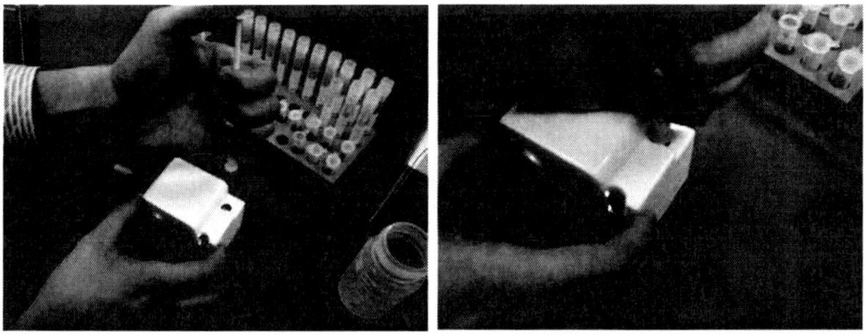

(a) (b)

Figure 5.34 Pictures of the sample injection on the third proto-model (a) preparation and (b) injection.

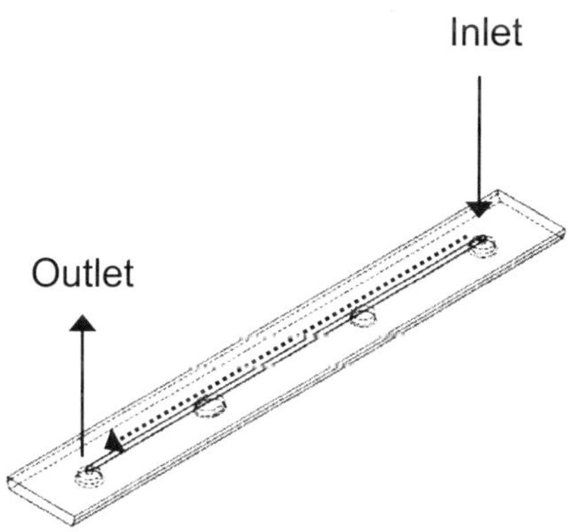

Figure 5.35 CAD image of the microchannel inside the sensor device.

5.3.3 Software

The calculation of the test result is carried out with a personal computer connected to the sensor system by USB cable. The electric power for the sensor system is also supplied through this USB cable. The time series data of the detection intensity is displayed for each array by the system software. The reflection coefficient can be coincidentally calculated by setting the reference intensity and background intensity in advance. The display image is shown in Fig. 5.36. To reduce the noise factors due to the fluctuations of the incident light and the transmittance of the sample solution, the detection intensity of fourth channel can be used as the compensation reference. When using this option, the nanopattern is not prepared on the array of the sensor device corresponding on the fourth channel. Therefore, the intensity from the fourth array does not include the reaction response but includes the fluctuations of reference light. By these contrivances, highly sensitive detection can be achieved with a very small system.

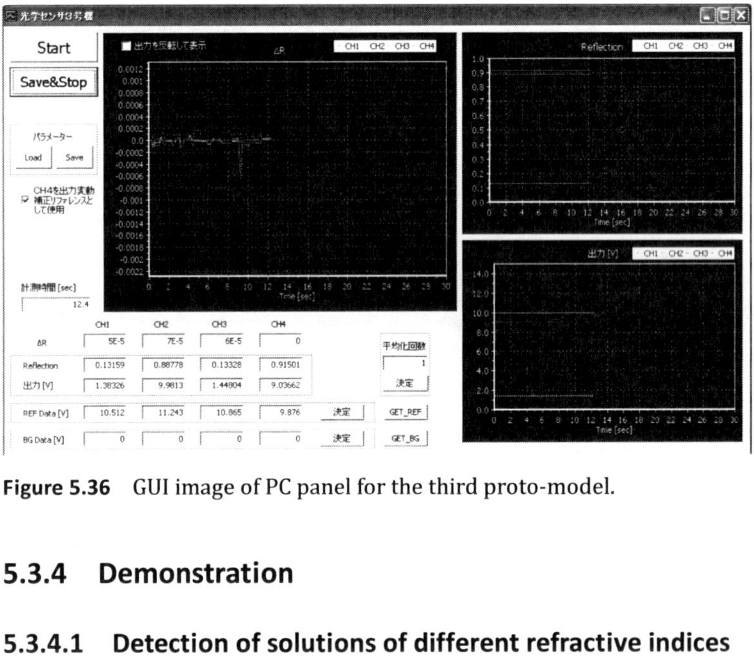

Figure 5.36 GUI image of PC panel for the third proto-model.

5.3.4 Demonstration

5.3.4.1 Detection of solutions of different refractive indices

As the evaluation of the constructed third proto-model, the sample solutions of the different refractive indices are detected. Five solutions whose refractive indices are 1.33, 1.36, 1.38, 1.40, and 1.43 are prepared by mixing purified water (n = 1.33) and ethylene glycol (n = 1.43) in different mixing ratios. They are sequentially injected into the sensor device by the syringe. In this experiment, the solution of n = 1.33 is injected after each sample solutions. The sequential data are shown in Fig. 5.37a. The signal shifts to the refractive index change are shown in Fig. 5.37b. This result verified that the signal function has a good linearity. Since the fabrication conditions of the sensor device are not optimized in this case, the error factors exist between the arrays. This will be improved by adjusting its fabrication conditions.

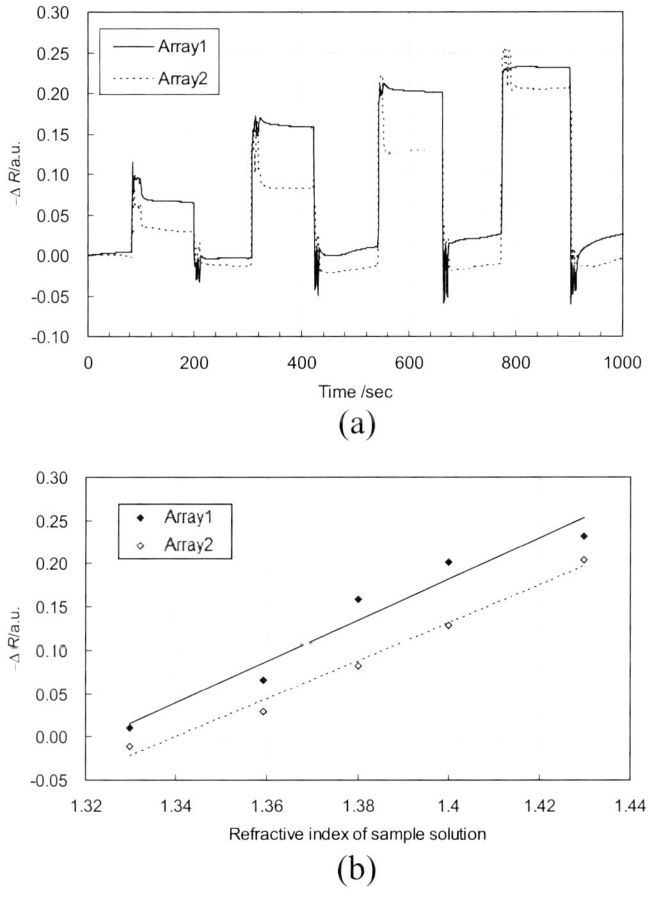

Figure 5.37 Experimental data demonstrating the basic characteristic of the third proto-model: (a) sequential data and (b) the signal shift to the refractive index of sample material.

In this experiment, it is proved that this system can basically work as a sensor to detect the change of the refractive index of the sample solution.

5.3.4.2 Protein detection

Subsequently, biomolecular interaction is detected by this system. The avidin protein is used as a model protein this time. The biotinylated polyethylene glycol-self-assembled monolayer (PEG-SAM) is formed by the following protocol as the first step.

(i) OEG-BAT (Biotin-terminated tri(ethylene glycol) hexadecanethiol, Asemblon, 381049-014) and HUT (11-Hydroxy-1-undecanethiol, Dojindo, H337) are dissolved in 99.5% ethanol. OEG-BAT solution (5% mol) is prepared by this process.

(ii) The above OEG-BAT solution is dropped on a bare gold surface of the sensor device.

(iii) After 10 min of immersion, the surface is rinsed by 99.5% ethanol.

(iv) The surface is dried out by a nitrogen gun.

After this preparation, the sample solution with the avidin protein (100 μg/mL) is injected. The experimental result is shown in Fig. 5.38. In this graph, the sample solution is injected at t = 50 sec. As a result, the reflection change of about 6% was observed. This result verified that the biomolecular interactions can also be detected by this palm-sized proto-model with a very simple and easy operation.

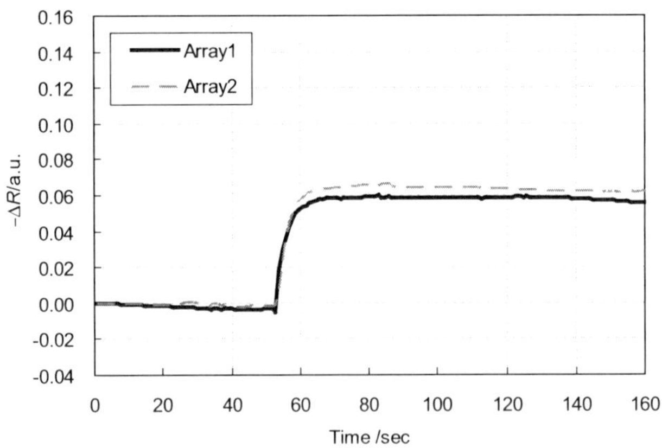

Figure 5.38 Experimental result of detecting avidin protein by using the third proto-model.

Chapter 6

Applications of Nanoimprint Biosensor

Heretofore, the concept, principle, device, and system of the nanoimprint biosensor have been mainly described. The experimental results and application examples of the nanoimprint biosensor are presented in this chapter.

6.1 Detection of α-Fetoprotein

Using the nanoimprint biosensor based on surface plasmon resonance (SPR) principle, we performed the quantitative detection of α-fetoprotein (AFP), a tumor marker. Although the AFP concentration in healthy human serum is approximately 20 ng/mL, its level increases markedly to more than several hundred ng/mL in patients with liver cancer (Teramura and Iwata, 2007). Currently, the cutoff value of AFP for clinical diagnosis is 200 ng/mL. Hence, the sensitive detection of the AFP using this system can be useful in cancer diagnosis.

6.1.1 Preparation of Probe Layer

To realize a high-sensitivity detection, the preparation of the probe layer is significant. Especially in nanoimprint biosensor, the thickness of the probe layer also needs to be taken into consideration. In this part, its preparation is presented. The experiment is conducted with the first proto-model which is introduced in Chapter 5.

Nanoimprint Biosensors: The Fusion of Nanofabrication, Nanophotonics, and Nanobiology
Takeo Nishikawa and Satoshi Fujita

ISBN 978-981-4316-67-5 (Hardcover), 978-981-4463-42-3 (eBook)
www.panstanford.com

For an SPR biosensing with nanoimprinted sensor device, the thickness of the probe layer to capture the target molecule should be smaller than 20 nm as the sensing depth is about 40–80 nm. To accomplish this purpose, we attempted to introduce an antibody-immobilization layer using ORLA18 protein (Orla Protein Technologies, UK). This ORLA18 protein is designed to present precisely oriented antibody (IgG)-binding domain structure as single layers (thickness ~10 nm) on surfaces (Athey et al., 2005). Its surface preparation process on the sensor device is explained below and depicted in Fig. 6.1.

(1) Treatment of the gold surface on nanoimprinting sensor device by injecting an aqueous solution containing 1% (v/v), β-mercaptoethanol.
(2) Self-assembly of the scaffold protein by injecting a 5 μM ORLA18 dissolved in Tris-HCl buffer (pH 8.0) containing *n*-octyl-β-D-glucopyranoside and ethylenediaminetetra-acetate (EDTA).
(3) Stabilization of the scaffold proteins and masking of the spaces between the proteins in the monolayer using filler molecule.
(4) Antibody binding on the ORLA18 protein layers by injecting 20 μg/mL antibody dissolved in Tris-buffered saline (TBS) (10 mM Tris-HCl, 150 mM NaCl, pH7.5).

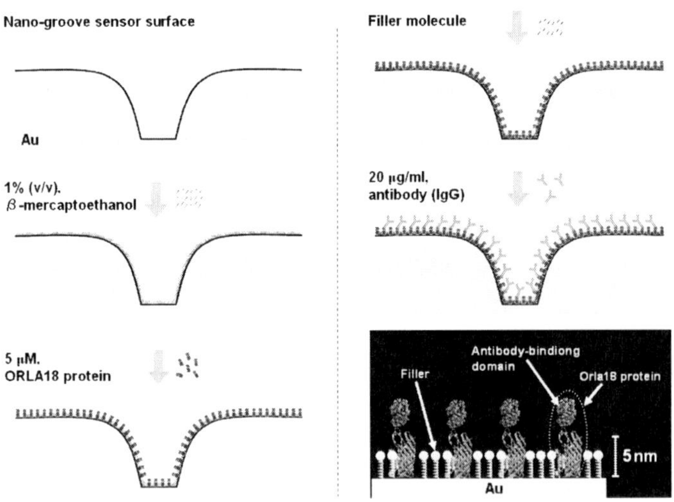

Figure 6.1 Process flow image of probe preparation on a nanostructured surface.

To ensure that the probe is firmly formed on this process, we have monitored the two-dimensional variation of the SPR peak wavelength shift during antibody-binding process. On the ORLA18 layer formed on the sensor surface, the solution containing the antibody is dropped in the left part of the sensor surface. The resonant wavelength changes before and after the antibody binding scanned in two-dimensional area. Figure 6.2 shows its result. Apparently, large signal change is observed only in the area the antibody is dropped. This result proved that the antibody is rightly prepared by the above immobilization process.

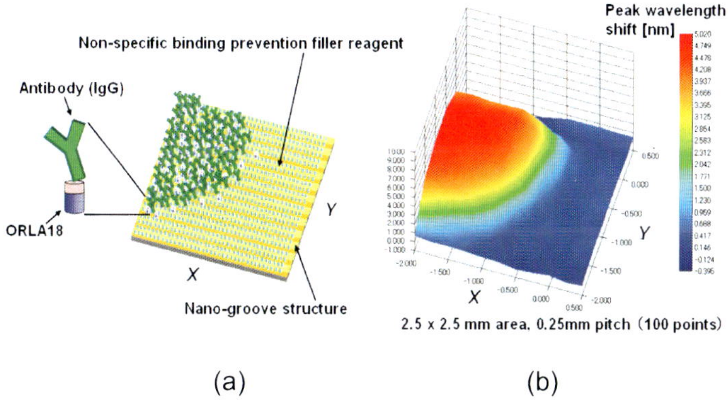

Figure 6.2 Two-dimensional monitoring of antibody binding on the ORLA18 surface using the nanoimprint biosensor system: (a) schematic image of antibody binding and (b) experimental result.

6.1.2 Direct Detection of AFP

As the first example, the direct detection of AFP is presented in this section. The direct detection means that AFP is detected without any modification on it before and after the injection. In general, direct detection is easy and simple though the sensitivity is not very high.

For the highly sensitive detection of AFP, an affinity-purified rabbit polyclonal antibody (95% IgG) against human AFP was purchased from Monosan (Uden, The Netherlands). Pure human AFP (a single band on SDS-PAGE) was obtained from Morinaga Institute of Biological Science (Yokohama, Japan). At first, we tried to detect the AFP molecules on the anti-AFP-immobilized ORLA18 surface at condition 1 in Fig. 6.3. However, the signal degradation was observed in low concentration. Then the detection limit of nanoimprinted SPR biosensor was estimated more than several hundred ng/mL (Fig. 6.4). After several experiments, we realized that to reduce the diffusion time of the AFP was essential. At the condition 1, mass-transport effect (Karlsson et al., 1991) becomes larger, because the size of its flow cell is larger. So, at first, we attempted to avoid mass-transport effect by high flow rate and large sample injection volume (condition 2). Figure 6.5 shows the result of AFP detection at condition 2. The error bar indicates three standard deviation of the base line. Finally, we performed decreasing the size of the flow cell (condition 3). At this condition 3, we could significantly improve the signal degradation in low concentration. As a result, we estimated that the detection limit of AFP by nanoimprinted SPR biosensor is approximately 20 ng/mL. This value already overcomes the cut-off value of 200 ng/mL in the clinical diagnosis.

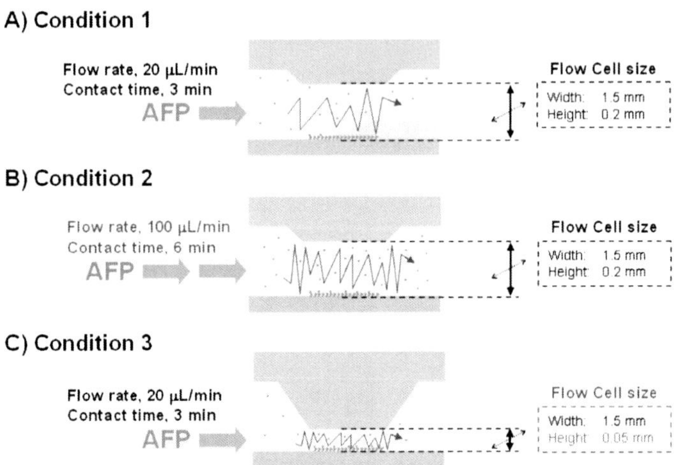

Figure 6.3 Experimental conditions of flow cell for AFP detection using the nanoimprint biosensor.

6.1.3 Signal Amplification by Gold Colloids

In case of other target proteins, the cut-off value can be much lower. To realize higher sensitivity, we have developed a signal amplification method by using gold colloids as a sandwich tool on a nanoimprint biosensor. Figure 6.6 shows the process image. On the sensor surface, ORLA18 was formed and the filler molecules were filled in its space to prevent the nonspecific binding. After that, anti-AFP antibody was prepared as the probe protein and the target protein AFP is sent over it. Then the biotinylated fragment of the anti-AFP antibody (Fab') was injected. Since this is a polyclonal antibody, it can bind to the AFP which exists on the sensor surface as a sandwich configuration. And as the last step, the gold colloids whose surface was coated by the streptavidin were sent on the sensor surface to amplify the signal.

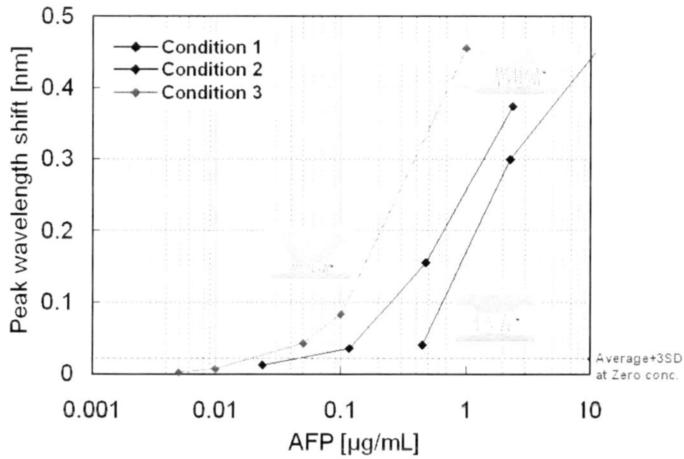

Figure 6.4 Experimental result for AFP detection on different conditions.

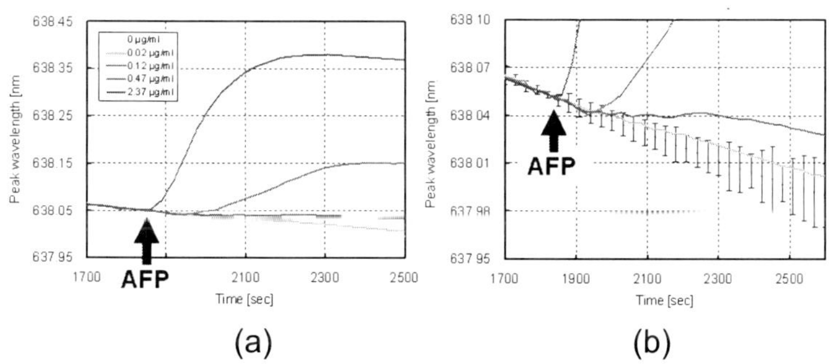

Figure 6.5 Time sequence experimental result of AFP detection.

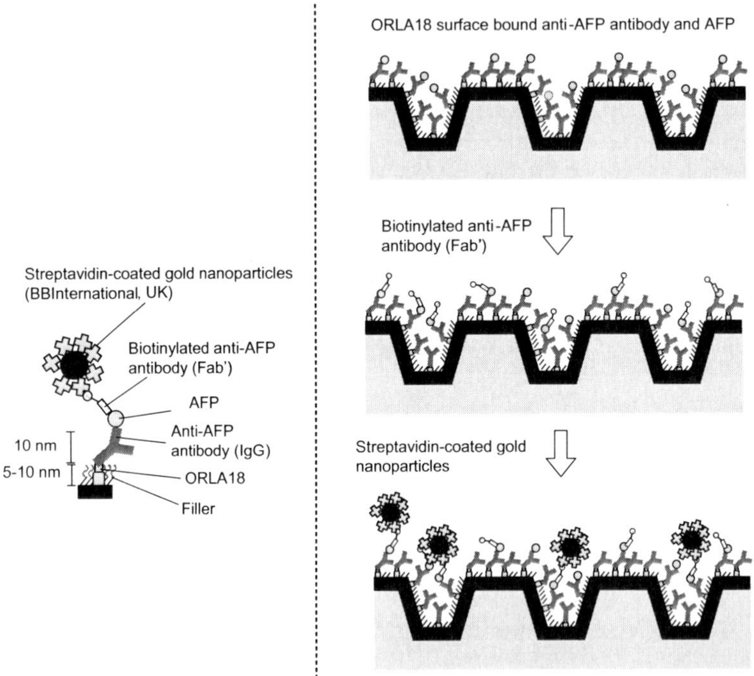

Figure 6.6 Schematics of the sandwich structure using gold colloid (left) and process flow (right).

6.1.3.1 Nonspecific binding evaluation of gold colloid

Since the nonspecific binding of the gold colloid can be a fatal problem, the nonspecific binding of the streptavidin-coated gold colloids on the sensor surface was testified in advance. The gold colloids commercialized from BBInternational (UK) were sent over the sensor surface on which ORLA18 and filler molecule are prepared. The diameter of the gold colloid was selected as from 2 to 40 nm. This experiment was conducted with the commercially available SPR equipment (Biacore-X, GE Healthcare, USA). The flow rate was 20 μL/min and HBS-EP (10 mM HEPES-NaOH, 150 mM NaCl, 3 mM EDTA, 0.005% (v/v) Tween 20, pH 7.4) was used as the buffer solution. The result is shown in Fig. 6.7. In case of the gold colloids whose diameter was 5, 20, and 40 nm, very large nonspecific binding was observed. On the other hand, small nonspecific binding was observed in case of 2, 10, and 15 nm. Especially, it was smallest in the gold colloid with 10 nm diameter. Since this result was reproducible, we had selected it for the signal amplification material. Although the correct reason why the quantity of the nonspecific binding depends on the diameter of the gold colloid is not clear, the surface of the gold colloids with the large nonspecific binding might have hydrophobic area in relation to the dimension of the streptavidin (about 5 nm).

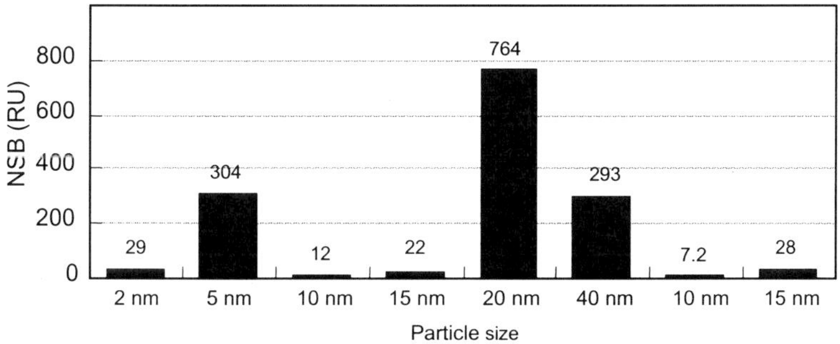

Figure 6.7 Nonspecific binding evaluation result on various colloid size conditions.

6.1.3.2 Signal amplification by gold colloids

The evaluation of the signal amplification with the above gold colloid is implemented on the nanoimprint biosensor. The experimental protocol given is below.

(i) ORLA18 surface preparation (in situ)
(ii) 5 μg/mL anti-AFP antibody (IgG)
(iii) 1 μg/mL AFP or 0 μg/mL AFP (buffer)
(iv) 10 μg/mL biotinylated anti-AFP antibody (Fab')
(v) 10% (v/v) streptavidin-coated gold colloid (10 nm)

The result is shown in Fig. 6.8. About the process (iii), 1 μg/mL AFP was injected in (a) and buffer solution was injected in (b) as a comparison. The data are depicted in Table 6.1. In the direct detection method of (a), the signal shift of 0.39 nm was observed. And further signal shift of 0.63 nm was observed by the injection of 10 μg/mL biotinylated anti-AFP antibody (Fab'). Then, the gold colloid was applied and the signal shift of 2.46 nm was added on it. On the other hand, almost no signal was observed through the same process in (b). These results mean that the signal was specifically amplified about 8 times as large as the direct detection method by using this sandwich method.

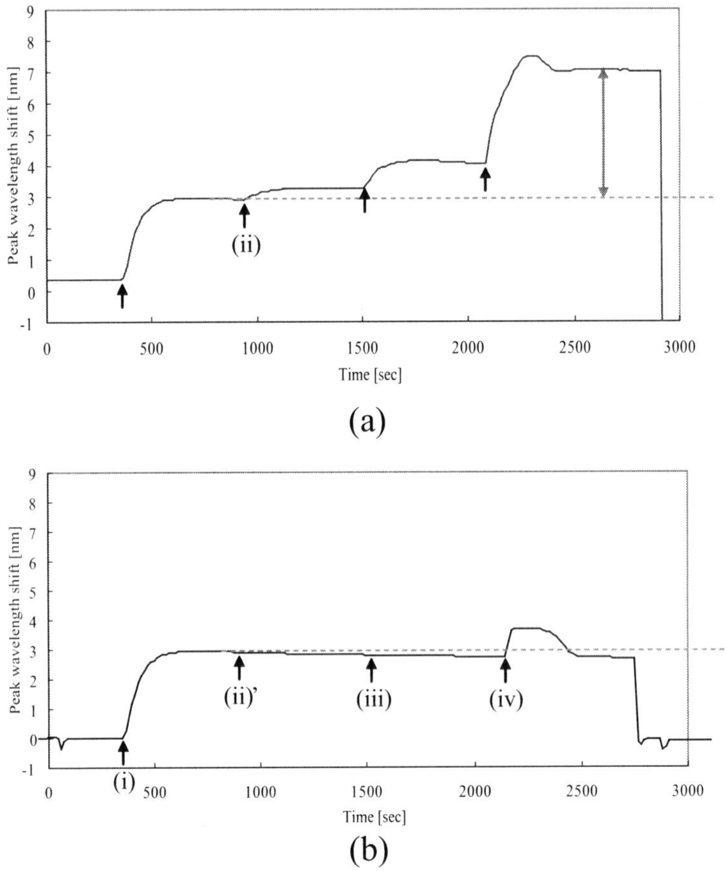

Figure 6.8 Experimental result of (a) signal amplification using gold colloid and (b) reference. (i) injection of 5 μg/mL anti-AFP antibody on the ORLA18 surface, (ii) 1 μg/mL AFP and (iii) buffer as a reference, (iv) 10 μg/mL biotinylated anti-AFP antibody, (v) 10% (v/v) streptavidin-coated gold nanoparticle (diameter, 10 nm).

Table 6.1 Signal shift of each process by the amplification method

	5 μg/mL anti-AFP antibody (IgG)	Ex1: 1 μg/mL AFP Ex2: buffer	10 μg/mL biotinated anti-AFP antibody (Fab')	10% (v/v) streptavidin-coated gold nanoparticles
Ex1. Positive	1.45 nm	0.39 nm	0.63 nm	2.46 nm
Ex2. Negative	1.69 nm	−0.10 nm	−0.05 nm	−0.03 nm

To grasp the distribution of the gold colloids on the sensor surface, the scanning electron microscopy (SEM) image of sensor surface after the injection of gold colloid was observed (Fig. 6.9). From this result, it is known that the colloids are immobilized on the sensor surface here and there. The density of the gold colloid on the flat area and the groove area is counted. The result is depicted in Fig. 6.10. According to this result, the density of the groove area is about 1.6 times as high as the one on the flat area. This seems to be caused by the increase of the surface area by the groove structure. By this result, it is proved that the gold colloid can intrude inside the groove structure and be immobilized there. This can provide the large signal amplification as shown in the above experiment.

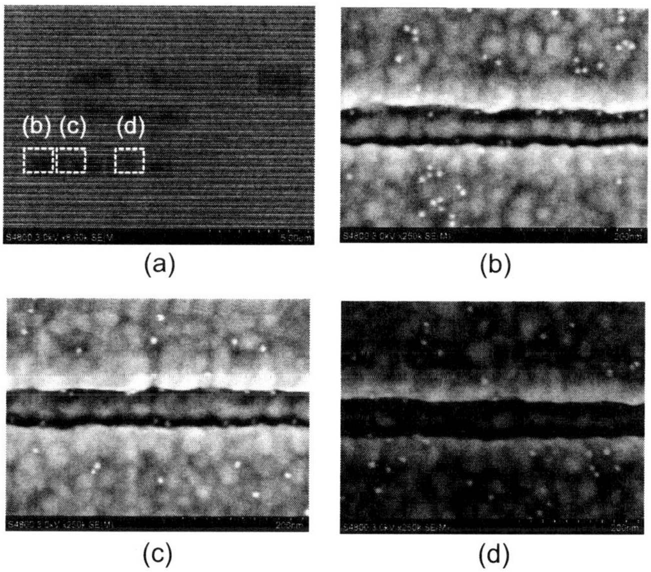

Figure 6.9 SEM images of the nanopatterns with the gold colloids (a) ×80,000 magnification and (b), (c), and (d) ×250,000 magnification.

In Fig. 6.11, the experimental result with relation to the concentration of AFP is exhibited. This result verified that the large signal amplification was realized in the low-concentration region, too. The detection limit of this experiment could be as small as 6 ng/mL by using gold colloidal signal amplification method.

By using these techniques, the nanoimprint biosensor can be applied to the wider applications.

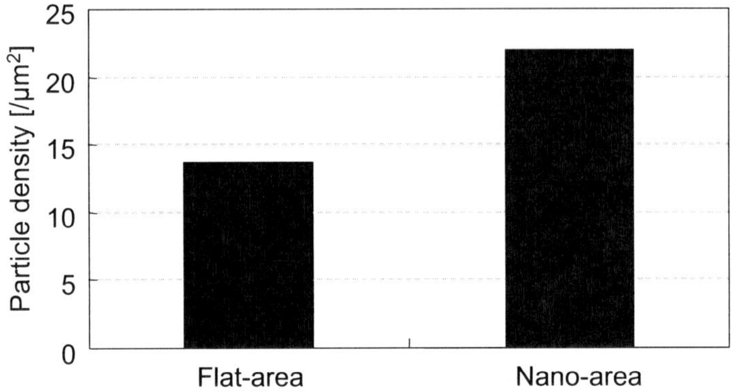

Figure 6.10 Gold colloid density counted from the SEM images on the flat area and concave area (nano-area).

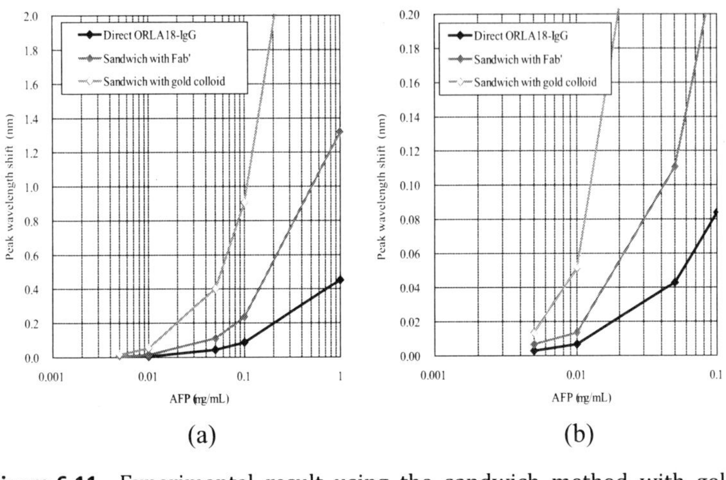

Figure 6.11 Experimental result using the sandwich method with gold colloid. (a) Calibration curve and (b) magnified graph in the low-concentration region.

6.2 Detection of Single-Stranded DNA

Recently, genetic diagnosis to predict the risk of cancer, lifestyle-related diseases, and other diseases is getting to be possible.

However, since it requires complicated and special equipments such as radioisotope labeling and gel electrophoresis to conduct DNA analysis, there is a need for more convenient and rapid detection method. In this study, the detection of a single-stranded DNA including the single nucleotide polymorphisms (SNPs) is demonstrated by using the nanoimprint biosensor. As no modifications of DNA sample such as fluorescent molecules and radioisotopes are necessary in this method, quick and easy detection and diagnosis of DNA can be possible.

6.2.1 Materials and Methods

As an experimental sample, single-stranded DNA related to the tumor necrosis factor (TNF) is synthesized and used. As the target DNA and SNPs, the base sequence DNA fragments of 5′-GGT TTC GAA GTG GTG GTC TTG-3′ (21 base), 5′-GGT TTC GAA GCG GTG GTC TTG-3′ (21 base) are prepared. And as the probe DNA, the complementary fragment to the target DNA (5′-biotin-AC CAC CAC TTC-3′) which is biotin-terminated to immobilize on the sensor surface is synthesized. As the first step, PEG-based self-assembled monolayer (SAM, biotin-terminated: no biotin-terminated = 1:99) is formed on the gold sensor surface. After that, the NeutrAvidin is bound and biotin-terminated probe DNA fragment is immobilized. On this prepared sensor, the target DNA, single base mismatch DNA, and random sequence DNA (as a reference) are injected (Fig. 6.12).

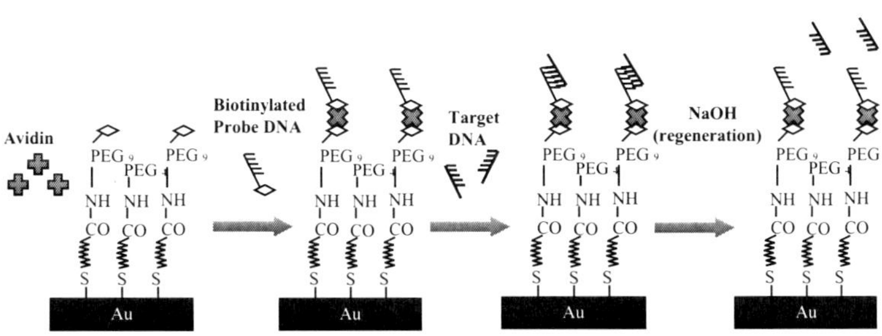

Figure 6.12 Schematic of DNA experimental protocol.

About the target DNA, the signal shift by changing its concentration is observed. To use the same sensor substrate continuously, the sensor surface is regenerated by the injection of NaOH. In this experiment, the nanoimprint LSPR sensor fabricated by the photopolymerization process is used. The pattern period and the gold thickness are 300 and 100 nm, respectively. The microchannel made of polydimethylsiloxane poly(dimethylsiloxane) (PDMS) whose width and depth are 2.0 mm and 25 μm is appressed against the sensor substrate. The flow rate was 40 μL/min. And the volume of injection sample was 200 μL.

6.2.2 Results and Discussions

The experimental result with the DNA sample of the different sequence injection is shown in Fig. 6.13. While almost no signal shift was observed in the random sequence DNA, large signal shift was observed in the target DNA fragment. Furthermore, a slight signal shift was observed when the SNP DNA was injected. In this experiment, it is verified that easy detection of target DNA fragment can be realized only in about 10 min by using the nanoimprint biosensor.

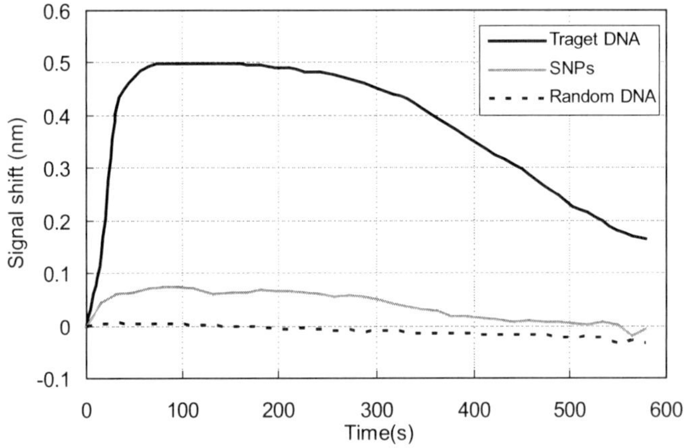

Figure 6.13 Time sequence experimental data after injecting the sample DNA.

In addition, the result of changing the sample concentration of target DNA is shown in Fig. 6.14. The signal shift at the time of 280 sec is shown in Fig. 6.15. These results mean that the signal shift value changes depending on the concentration of target DNA.

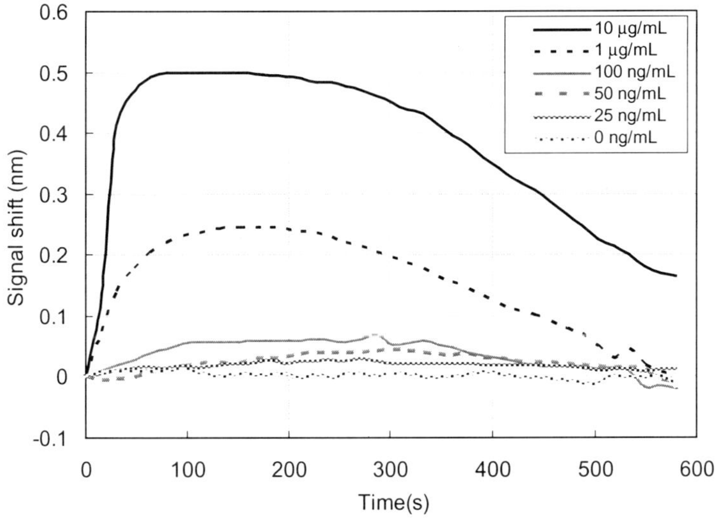

Figure 6.14 Time sequence experimental data with changing the concentration of the target DNA.

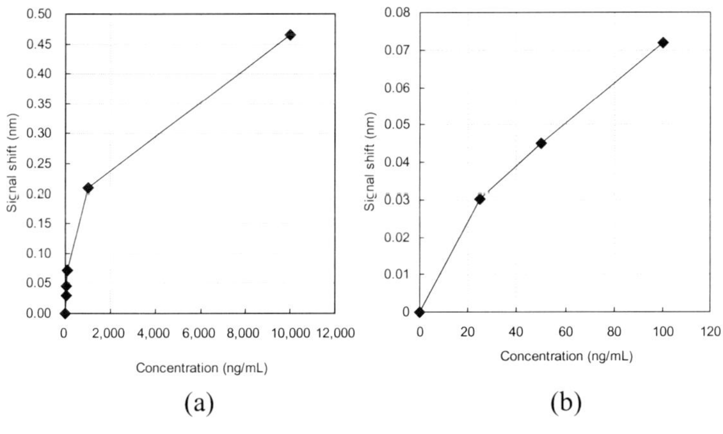

Figure 6.15 Calibration curve of target DNA detection: (a) calibration curve of target DNA detection, and (b) magnified graph in low-concentration region.

From these results, not only protein molecules but also DNA fragment including SNPs can be easily detected by the nanoimprint biosensor. By using this sensor, medical diagnosis based on genetic information can be realized in easier way and in low cost.

6.3 Signal Amplification Method by Magnetic Beads

Magnetic beads have been used for the separation of blood components in the sample and so on in the biochemical tests. In recent years, its development is keenly progressed and the ones with smaller diameter are commercialized. In the conventional SPR, the use of the magnetic beads has been limited since the optical prism is located at the backside of the sensor device. On the other hand, the magnet for attracting the magnetic beads can be directly attached at the backside of the nanoimprint LSPR device since it does not have the prism. In this study, the possibility of much higher sensitive detection is examined by using the magnetic beads in the nanoimprint biosensor.

6.3.1 Basic Concept

Our previous study proved that the detection limit can be much improved by using gold colloids. However, the detection of the very low concentration target such as less than 10 pg/mL is still challenging even with these modification methods. As an aforethought, the amount of the protein contained in the sample is considered here. In the 100 μL amount of the sample solution with AFP (70 kDa) of 10 pg/mL concentration, 1 pg AFP exists in the sample solution in total. The signal change when the total AFP is supposed to bind to the sensing spot whose diameter is 300 μm (about 0.07 mm^2) is as small as about 14 RU (Resonance Unit) since the reaction of 1 ng/mm^2 corresponds to 1000 RU in SPR. This calculation result means that the target molecules should bind to the sensing spot with as small loss as possible to detect the very low concentration target solution. In this behalf, the loss factors of the target molecule in the present SPR system are described below.

(i) **Captured loss of the target molecules on the upstream of the sensing spot.**

The target molecules can be captured at the inner wall of injector, mirochannel by the nonspecific binding. To reduce it, the selection of the surface material and the coating of inner surface to prevent the nonspecific binding are important. Furthermore, the target molecules can bind to the probe immobilized on the upstream of the sensing spot by the specific binding. To reduce it, the probe materials should be immobilized only on the sensing spot by using the spotting technique and so forth.

(ii) No contact of the target molecules with the sensing spot.

If the target molecules does not contact with the sensor surface on which the probe molecules are immobilized, no biding can occur. The detail will be described below. When the contaminants such as blood cells and other biomolecules exist in the sample, the possibility of the target molecule to contact with the sensor probe can be reduced.

(iii) No binding of the target molecule with the probe molecule even when its contact occurs.

When the immobilization of the probe molecule is not right, the target molecule cannot bind with them. The orientation, density, etc. should be optimized to improve the binding efficiency. The association constant of the probe molecule should be large to realize the strong affinity with the target molecule.

(iv) Dissociation of the target molecule after the association reaction.

When the dissociation constant is large, the target molecule tends to dissociate from the probe molecule after the association reaction. The selection of the probe molecule, immobilization method, and buffer conditions (pH, surfactant, etc.) should be necessary.

Since the improvements about (i), (iii), and (iv) have been introduced in the previous chapters, the details of (ii) will be considered here. Figure 6.16 shows the magnified image in the sample solution on the sensor surface. In the case that the concentration of AFP is 10 pg/mL, about 86 million molecules exist in 1 mL sample solution. And the distance to the neighboring molecule is about 23 μm which is very far considering that the size of AFP molecule is only about 10 nm. The molecules in the sample solution proceed 66.7 mm in a second when the flow velocity is 40 μL/min and the cross-sectional dimension of the microchannel is 1 mm width and 10 μm height. When the length of sensing region along the micro-channel is 300 μm, it takes only 0.0045 sec for the molecules in the solution to cross over the sensing region.

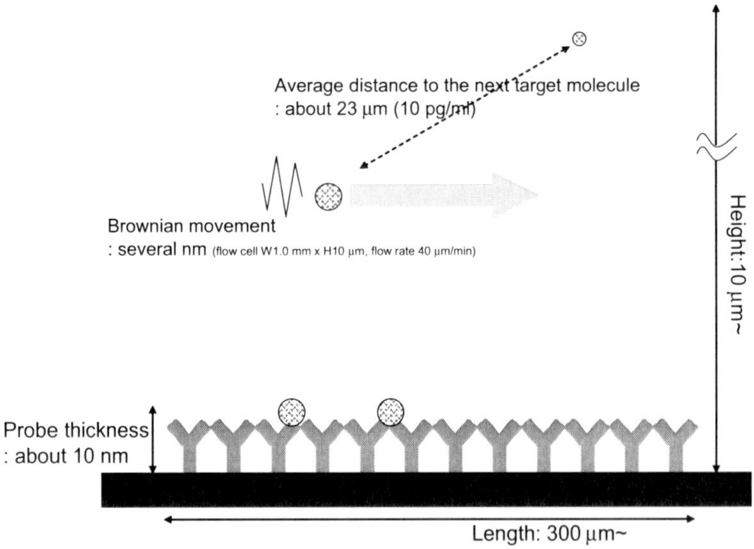

Figure 6.16 Schematic image of flowing proteins inside the flow cell.

According to Einstein's theory of Brownian motion and diffusion, the average moving distance in time "*t*" can be expressed as the following equation.

$$x = \sqrt{\frac{RTt}{3\pi\eta a N_A}}$$

where R is the gas constant, N_A is Avogadro's constant, η is the viscosity coefficient, and a is the radius of the molecule. In purified water at 20°C, a molecule with 5 nm radius moves 9.3 μm in a second in average according to this theory. Therefore, the molecule moves only about 0.62 μm while passing through over the sensing spot (0.0045 sec). This means that the diffusion distance of molecules by Brownian motion is much smaller than the dimension of the microchannel. Since the thickness of the probe layer is about 10 nm and the height of the microchannel is 10 μm, most of the target molecules (more than 99%) might pass through over the sensor surface without contacting to the probe molecules according to these results. To improve the detection limit of a sensor, the ratio of target molecules contacting to the probe molecules needs to be enlarged since the amount of the target molecules is inherently very small in a low-concentration (such as 10 pg/mL) sample solution as described above. By making the height of microchannel thinner, the possibility of target molecules to contact to the probe layer gets larger theoretically. However, it is practically difficult to make it thinner than 10 μm because the pressure loss at the microchannel also gets remarkably large. And the microchannel can get easy to be blocked up when it gets narrower and thinner.

We have, therefore, proposed a unique method to force the target molecules flowing over the sensor surface to be pulled down to the sensor surface by using the magnetic beads (Fig. 6.17). The basic protocol is shown in Fig. 6.18. The probe molecules such as antibody are immobilized on the sphere surface of the magnetic beads in advance. And after mixing the magnetic beads with the sample solution, the mixed solution is sent over the sensor surface. On the surface of the sensor substrate, the probe molecules are immobilized, too. Under the sensor substrate, a magnet is prepared. The magnetic beads are forced to be attracted down to the sensor surface by the magnetic force. Then, by releasing the magnet under the sensor substrate, the magnetic beads capturing the target molecules remain on the sensor surface due to the binding interaction with the probe molecules on the sensor surface. On the other hand, the magnetic beads without the target molecules are released from the sensor surface by the flow force in the microchannel. By this procedure, the target molecules which have been passing over the sensor surface without contacting to the probe molecules can be forced to approach to the sensor surface by the magnetic force. And the signal amplification is also expected since the size of the magnetic beads is much larger than the size of a target molecule. It is theoretically expected that the detection limit of the sensor is improved by this method.

6.3.2 Materials and Methods

To demonstrate the experiment based on the above concept, the preparations are presented in this subsection.

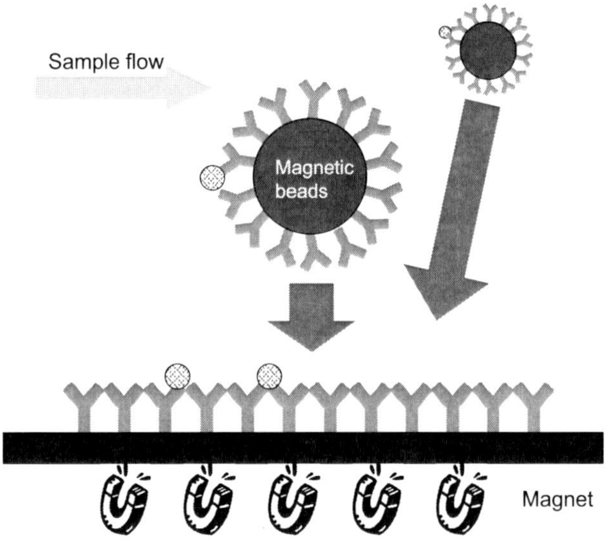

Figure 6.17 Schematic of new sensing technology to improve the detection limit by using magnetic beads.

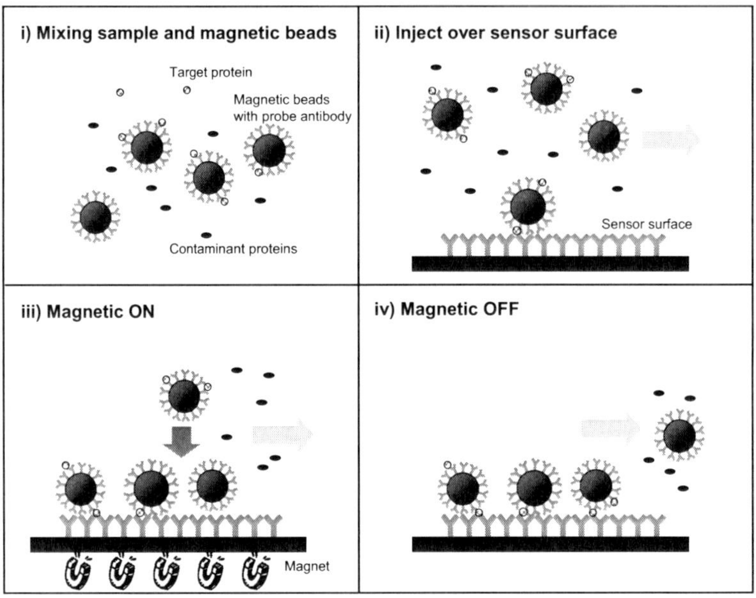

Figure 6.18 Process steps using the magnetic beads.

6.3.2.1 Selection of magnetic beads

In recent years, various types of magnetic beads are commercialized. About the diameter, the larger it is, the stronger the effect by the magnetic force is. However, the reaction coefficient goes down when it gets larger. And when its diameter is too small, it takes very long time to pull the magnetic beads in a liquid solution. To achieve a high sensitivity with the magnetic beads, the selection of the appropriate diameter is necessary. We have evaluated some of the commercially available magnetic beads in this study.

6.3.2.2 Probe immobilization on the magnetic beads

To realize this protocol, the antibody needs to be immobilized on the magnetic beads in advance. Figure 6.19 shows its process flow. For preventing the aggregation of the magnetic beads when the target molecule is mixed, monoclonal antibody is used herein. As the experimental condition for binding the antibody, four conditions shown in Table 6.2 are studied. A spectrometer is used for measuring the quantity of immobilized antibody on the magnetic beads (Fig. 6.20). The absorbance intensity of the sample solution containing the antibody at the wavelength of 280 nm (A_{pre}) is measured before mixing it with the magnetic beads. After the mixing, it is measured again (A_{post}) and the change ratio is calculated by the following equation. The larger the change ratio is, the more quantity of the antibody is immobilized on the magnetic beads.

$$\frac{A_{pre} - A_{post}}{A_{pre}} \times 100$$

Table 6.2 Four experimental conditions for immobilizing antibody on the magnetic beads

	Fc1	Fc2	Fc3	Fc4
Immobilized antibody	Anti-I antibody	Anti-I antibody	Anti-L antibody	Anti-L antibody
Antibody concentration	20 μg/mL	100 μg/mL	20 g/mL	100 g/mL
Buffer pH	5.0	6.3	5.0	6.3
Buffer Solution	50 mM NHS 200 mM EDC	17 mM NHS 5 mM EDC	50 mM NHS 200 mM EDC	17 mM NHS 5 mM EDC

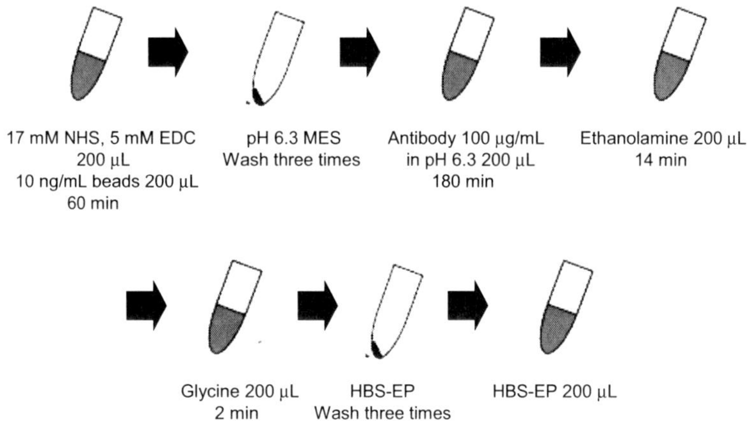

Figure 6.19 Schematic of immobilization protocol of antibody on the magnetic beads.

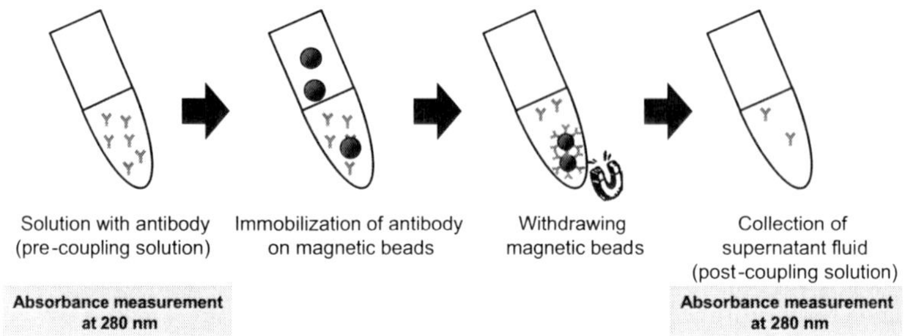

Figure 6.20 Quantity measurement method of the antibody immobilized on the magnetic beads.

The result is shown in Table 6.3. It is proved that the quantity of the bound antibody is most on the experiment condition of Fc1. The following experiment is conducted with the magnetic beads on which the antibody is immobilized by this process.

Table 6.3 Experimental result of absorbance measurement before and after immobilizing antibody on the magnetic beads

	Fc1	Fc2	Fc3	Fc4
Pre-CS [a.u.]	0.041	0.213	0.025	0.147
Post-CS [a.u.]	0.023	0.133	0.020	0.098
Difference [%]	43.9	37.5	20.0	33.3

6.3.3 System Construction

To attract the magnetic beads flowing in the microchannel down to the sensor surface, a magnet needs to be located at the backside of the sensor substrate.

6.3.3.1 Selection of the system conditions

To realize the experimental system based on this concept, the magnet needs to be switched on and off arbitrarily. The electromagnet can be switched only by controlling the electric current without any mechanical motion. However, the magnetic force of the permanent magnet made of the rare-earth material is generally stronger than that of electromagnet when the space is limited. In this part, we have selected the magnet type by comparing the electromagnet and the permanent magnet.

Figure 6.21 shows the schematic image of the experimental setup to evaluate the trapping of the magnetic beads in the liquid flow by the magnet. The back plate of the sensor device holder is hollowed out to approximate the magnet to the sensor substrate. The distance between the sensor substrate and the top of magnet can be less than 1 mm by this system. A microchannel made of PDMS is interleaved between two transparent glass substrates. To observe the state of the magnetic beads apparently, a white paper is put under the sensor substrate and a digital camera is set over the sensor substrate to take its pictures.

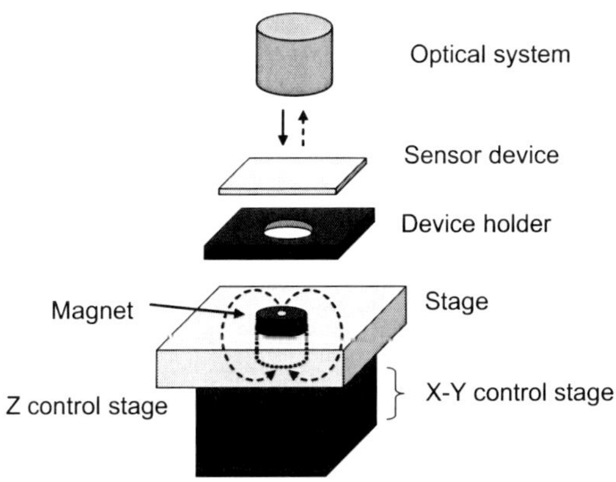

Figure 6.21 Schematic of the experimental setup for magnetic beads experiment.

The width and depth of the microchannel are 2 mm and 50 μm, respectively. And the purified water was sent by the syringe pump (PHD2000, Harvard Inc., USA). The magnetic beads solution of 100 μL volume was injected by a manual injector.

As the magnet, an electromagnet (ϕ15 mm × L15 mm, rated voltage 35 V, TMN-158S, Gigateco Co. Inc., Japan) and a neodymium magnet (ϕ6 mm × L10 mm, Sangyo Supply Co., Japan) are testified. As the magnetic beads, Dynameads (1 μm diameter) and nanomag-D (50, 130, and 250 nm diameter) were prepared. The flow rate was set at 40 μL /min.

The experimental result is shown in Fig. 6.22. The sample flows from right side to left side. The magnet is located at about the center of the picture. In the result of electromagnet, it is observed that the magnet beads are trapped at the downstream of the magnet when their diameter is 1 μm. However, when their diameter is 250 nm, they are not trapped by the electromagnet. While the pressure from the sample flow decreases by downsizing the magnetic beads,

Magnet	Diameter of magnetic beads	Flow rate			Results
		40 μL/min	200 μL/min	500 μL/min	
Neodymium (ϕ6 mm)	D1.0 μm (12-fold dilution)	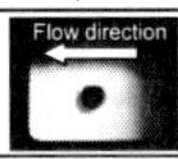		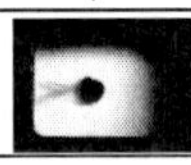	Trap and release of magnetic beads by the magnetic force are OK.
	D 250 nm (100-fold dilution)	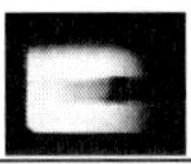			In low flow rate, trap is OK.
	D 130 nm (30-fold dilution)				In low flow rate, trap is OK.
	D 50 nm (10-fold dilution)				Cannot trap the magnetic beads.
Electromagnet (ϕ15 mm,35 V)	D 1.0 μm (12-fold dilution)				Cannot trap the magnetic beads.
	D 250 nm (100-fold dilution)	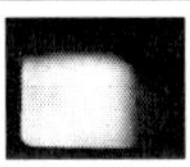			Cannot trap the magnetic beads.

Figure 6.22 Observation of trapping the magnetic beads by the magnet on various experimental conditions.

the pulling force from the magnet also decreases by their downsizing. It results in that the magnetic beads of 250 nm diameter cannot be trapped by the electromagnet. By enlarging the size of electromagnet, the magnetic force can be enhanced. In the actual experimental setup, however, the size of the sensing spot is limited and the target molecules trapped at the outside of the sensing spot does not contribute to the signal change. Therefore, the improvement by using larger electromagnet is expected to be no practical solution for this experiment.

In the experimental result using neodymium magnet, the magnetic beads of 1 μm diameter are trapped at about the center of the magnet location. In the case that the magnetic beads of 250 nm, 130 nm are used, they are also trapped by the neodymium magnet. About the magnetic beads of 50 nm, however, they were not trapped even by the neodymium magnet. These results verified that the magnetic force to pull down the magnet beads of the neodymium magnet is larger than that of the electromagnet. And the magnetic beads whose diameter is larger than 130 nm need to be used to realize the above concept.

6.3.3.2 Construction of the experimental setup

From the above experimental investigation, the neodymium magnet needs to be located under the sensor device. When using the permanent magnet, it cannot be switched on and off electrically. Therefore, the switching is achieved by controlling the distance between the magnet and the sensor device in a mechanical fashion. As the first trial, we have constructed an experimental system in which the magnetic force is controlled by inserting and removing the neodymium magnet under the sensor substrate, as shown in Fig. 6.23. To achieve a high-sensitivity detection, the noise element by the effect of switching the magnetic force should be very small. To verify this effect, the nanoimprint SPR sensor device was placed on the chip holder. And the neodymium magnet was switched on and off while the resonant wavelength was observed. As the result of this case, the noise fluctuations occurred at the switching points of magnet as shown in Fig. 6.24. This came from the slight change of the distance between the object lens and sensor device due to the chip holder weight's ups and downs.

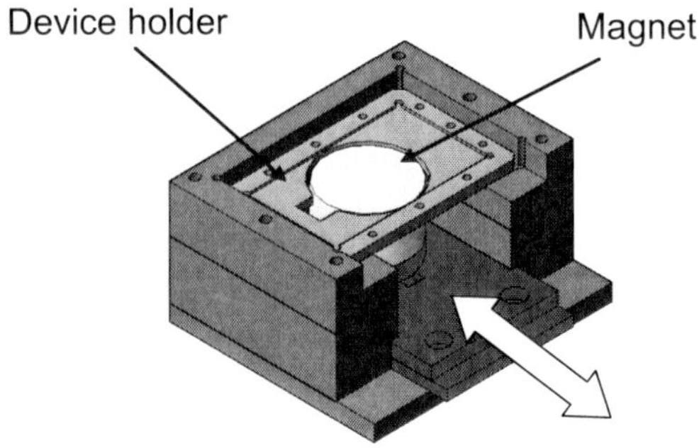

Figure 6.23 CAD image of the device stage to insert and remove the magnet under the sensor device.

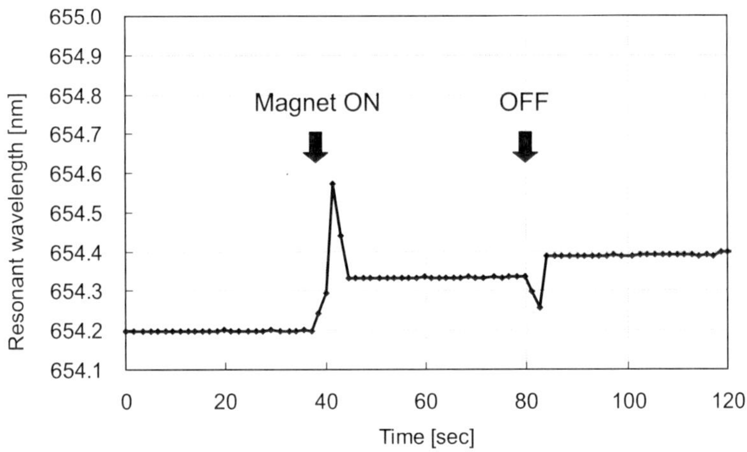

Figure 6.24 Experimental result to evaluate the stability of the magnet system in which the magnet is inserted (on) and removed (off).

Then, the system in which the magnet is not removed completely but is rotated to switch the magnetic force on and off is constructed as an improved version of the system (Fig. 6.25). In this system, the angle of a base plate holding the magnets is controlled by rotating a handle attached at a side of the sensor holder. The experimental result of similar verification is shown in Fig. 6.26. In comparison with the previous system in which the magnet is removed and inserted, noise hardly occurred and an excellent result was obtained. This improved system is used to verify the possibilities of detection limit improvement by controlling the magnetic beads.

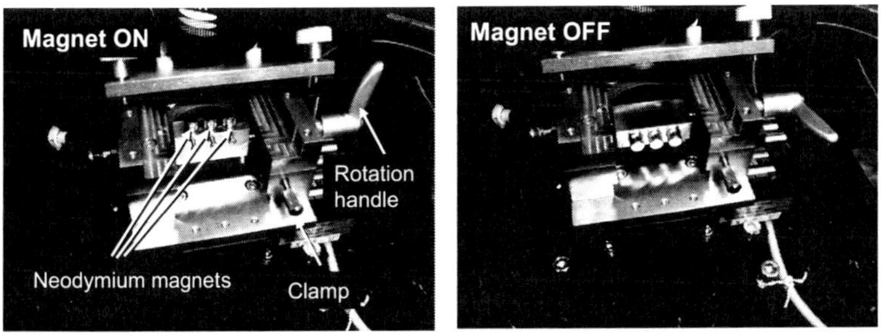

(a) (b)

Figure 6.25 Improved experiment system to turn the magnet on and off under the sensor device by rotating it.

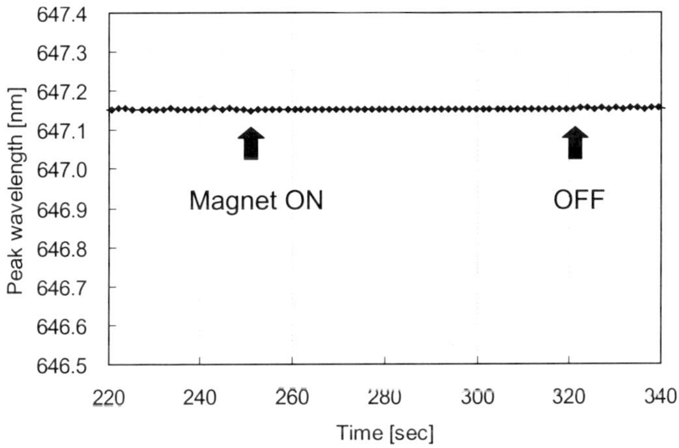

Figure 6.26 Stability evaluation of the improved experimental system for turning the magnet on and off.

6.3.4 Nonspecific Binding of Magnetic Beads on Sensor Surface

To achieve high sensitivity by using the magnetic beads, its nonspecific binding needs to be as small as possible. In this subsection, the nonspecific binding of the magnetic beads on the sensor surface is investigated in advance. The nanoimprint sensor device with the nanogroove structure of 300 nm period fabricated by photopolymerization process is used here.

6.3.4.1 Bare gold surface and purified water dilution

As the first experiment, the signal change caused by the nonspecific binding of the magnetic beads diluted by the purified water on the bare gold sensor surface is testified. The magnetic beads of 130 nm diameter coated with PEG-COOH (nanomag-D, Micromod, Germany) are used with being diluted 300 fold ("x" in the graph) and 30 fold ("y" in the graph) by the diluted water. The volume of the injection loop was 100 μL and the width and depth of the microchannel made of PDMS were 2.0 mm and 25 μm, respectively. The flow rate of the sample solution was 40 μL/min. And the neodymium magnet with 6 mm diameter is used.

The experimental result observing the resonant wavelength of LSPR with sample flowing is shown in Fig. 6.27. At about t =200 sec, ethanol is injected to wash out the surface of the bare gold. And at about t = 950 sec, the sample solution of (x) was injected four times while the magnet is off. The signal change of about 1 nm resonant wavelength shift was observed at each injection time. Subsequently, the sample solution of (y) is injected four times while the magnet is off. The signal changes of as much as about 10 nm were observed due to the first, second, and third injections, respectively. Only small signal change was, however, observed by the fourth injection due to the signal saturation. After that, though the sample solution of (x) is injected with the magnet on at about 5600 and 7100 sec, almost no signal change could be observed. Since the large signal changes have been observed even during the off period of the magnet, these signal changes seem to be caused by the nonspecific binding of the magnetic beads on the bare gold surface. The picture of the sensor chip after this experiment is shown in Fig. 6.28a. The dark line was apparently observed in concurrence with the attached place of microchannel. The SEM image of the area on which the magnetic beads is flown is shown Fig. 6.28b. It is observed that the magnetic beads are covering the most part of the sensor surface. Some of them fill the nanogroove structure on the sensor surface by the nonspecific binding. This result verified that the magnetic beads diluted by the purified water easily bind to the bare gold surface even if their surface is coated by the PEG-COOH.

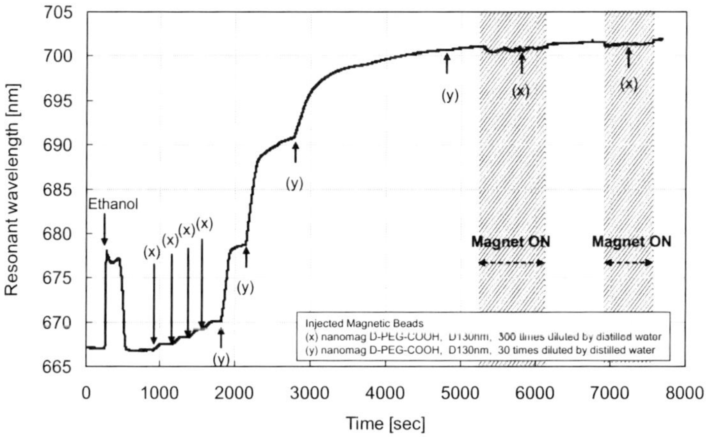

Figure 6.27 Evaluation result of the nonspecific binding of magnetic beads on the bare gold sensor surface.

Figure 6.28 The sensor device evaluation after the nonspecific binding experiment: (a) a picture of sensor device and (b) SEM image at the nanopatterned area inside the microchannel.

6.3.4.2 PEG-coated gold surface and buffer dilution with surface detergent

As the second experiment to suppress the nonspecific binding of the magnetic beads, the sensor surface whose surface is coated with PEG is used. Furthermore, the magnetic beads are diluted by HBS-EP buffer with the surface detergent. Other experimental parameters were basically same as the previous experiment.

Figure 6.29 shows the result with the sample (x). At about 700 and 1200 sec, the magnetic beads were injected. No apparent signal shift as the previous experiment was observed, though some fluctuation noises due to the floating magnetic beads exist during the sample flow. And then the magnetic beads were injected at about 2150 sec while the magnet is on. During the "on" period of the magnetic force, large signal change was observed caused by the accumulation of the magnetic beads. However, the signal recovers to the initial position by turning off the magnet. The same experiment was conducted with the sample (y) as shown in Fig. 6.30. While the signal largely changed with the magnet on, it recovered to the initial point with the magnet off, too. These results proved that the nonspecific binding of the magnetic beads can be prevented by coating the sensor surface with PEG and by diluting the magnetic beads with the buffer solution including the surface detergent.

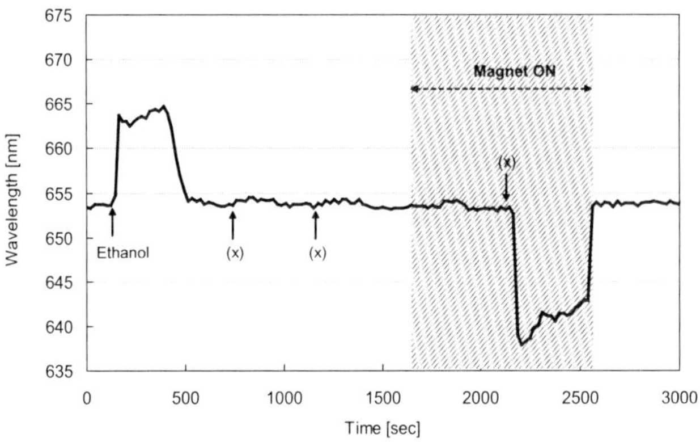

Figure 6.29 Nonspecific binding evaluation result of the magnetic beads on the surface modified sensor device.

When the magnet is on, the signal shifts to the minus direction in the experiment with sample (x) and it shifts to the plus direction in the one with sample (y). To analyze these results, the reflection spectra with (x) and (y) are investigated. In Fig. 6.31, the reflection spectra at t =750 sec (magnet is off), t = 2413 sec (magnet is on with sample (x)), and t = 16014 sec (magnet is on with sample (y)) are exhibited. When magnet is off, the resonant dip is observed at about 650 nm. At t = 2413 sec, the reflection spectrum totally decreases and the spectrum curve gets very broad. This is caused because the aggregation of the magnetic beads covers the sensor surface and the reflection intensity gets very small. At t = 16014 sec, the reflection spectrum almost becomes 0 in all wavelengths. In these cases, the apparent reflection dip does not exist. Therefore, the shift results of resonant wavelength with the magnet on are meaningless. Since the important result in this experiment is the shift value of the resonant wavelength when the magnet is off again after the on period, the effect of the aggregation of the magnetic beads with the magnet on does not matter too much.

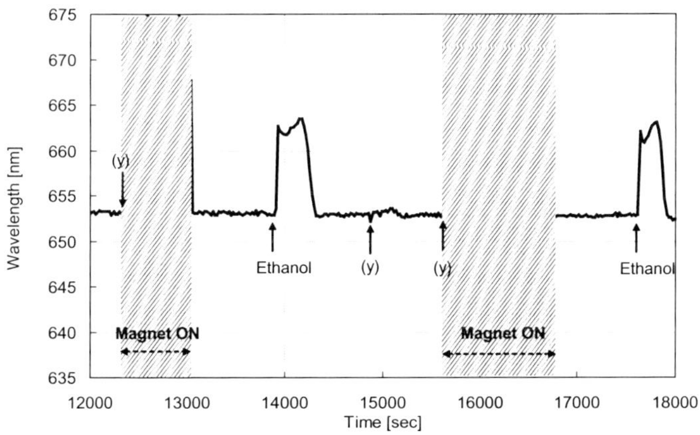

Figure 6.30 Nonspecific binding evaluation result of the magnetic beads on the surface-modified sensor device with sample (y).

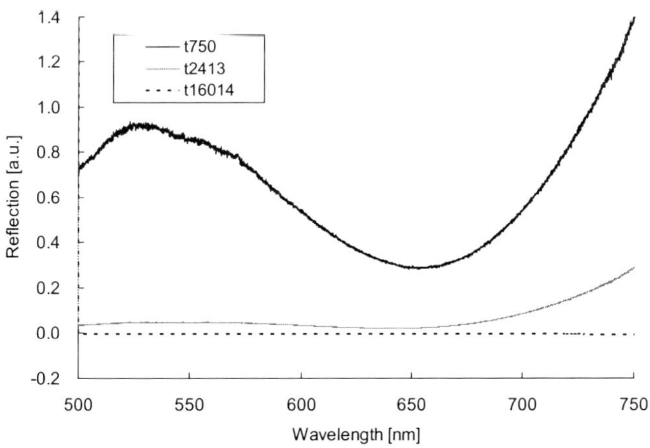

Figure 6.31 Reflection spectrum of the nonspecific experiment data.

6.3.5 Improvement of Detection Limit

6.3.5.1 Preparation of experiment

Subsequently, on the experimental condition as described above, the possibility of the sensitivity improvement is investigated by using the magnetic beads. The AFP is detected by a sandwich method using the magnetic beads bound on the anti-AFP antibody. The experimental parameters such as the size of microchannel, the volume of the sample injector, and flow rate are same as the previous experiment.

As the preparation of the experiment, anti-AFP antibody was immobilized on the sensor surface as the probe layer. The sensor diagram is shown in Fig. 6.32. In advance, the PEG-SAM layer was prepared on the sensor surface by overnight dip method. *N*-hydroxysuccinimide/ethyl(dimethylaminopropyl) carbodiimide (NHS/EDC) was injected to activate the surface of PEG. And then anti-AFP antibody of 20 μg/mL was injected. The activated surface was blocked by ethanol amine after that. And the surface was washed by glycine HCl. In this experiment, each reagent was injected two times for assurance. Finally, the signal shift of about 1.5 nm was observed due to the immobilization of antibody. It was also verified that nonspecific binding of magnetic beads did not occur with the magnet on and off, though the result is not shown in this graph.

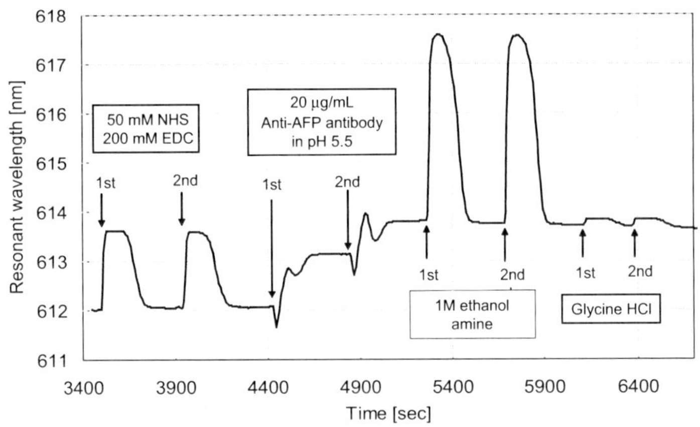

Figure 6.32 Sensor diagram of the immobilization of anti-AFP antibody as the probe protein.

By using this surface, two experiments shown in Fig. 6.33 were carried out.

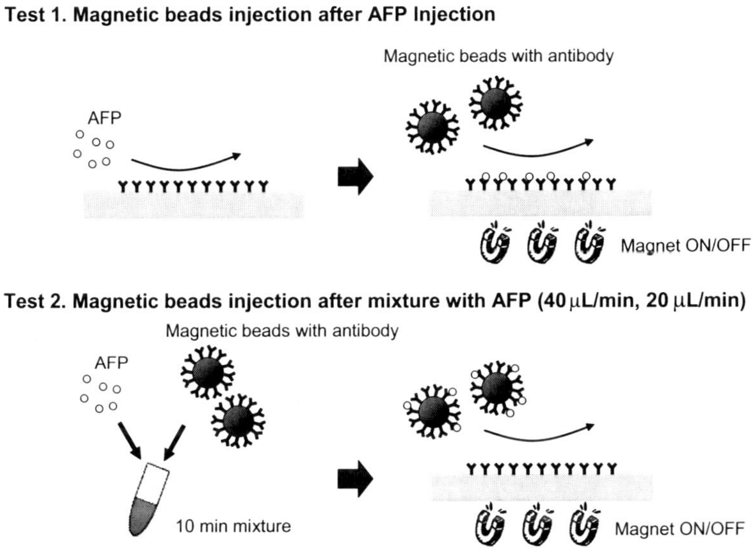

Figure 6.33 Schematics of the two experiments using magnetic beads and nanoimprint SPR sensor.

In the first experiment, the magnetic beads on which the anti-AFP antibody was bound were sent over the sensor surface after the injection of AFP (1 μg/mL). Under both conditions of magnetic force (on and off) during the flow of magnetic beads, this experiment was conducted and the effect of magnetic force to pull the magnetic beads was verified. In the second experiment, AFP and the magnetic beads on which anti-AFP was bound were mixed for 10 min in advance. And then, this mixture was sent over the sensor surface. This experiment was also conducted under both conditions of magnetic force (on and off). And the effect of the flow rate is also testified. The flow rate was set at 20 and 40 μL/min (typical value for other experiments). The magnetic beads used in these experiments was nanomag-D (D130 nm, surface:PEG-COOH, 10 mg/mL in PBS buffer pH 7.4) on which anti-AFP antibody was bound. The stock solution was diluted 10 fold by HBS-EP.

6.3.5.2 Magnetic beads injection after AFP injection

In this experiment, the magnetic beads solution is injected after the injection of AFP. The sensorgram result is shown in Fig. 6.34. In

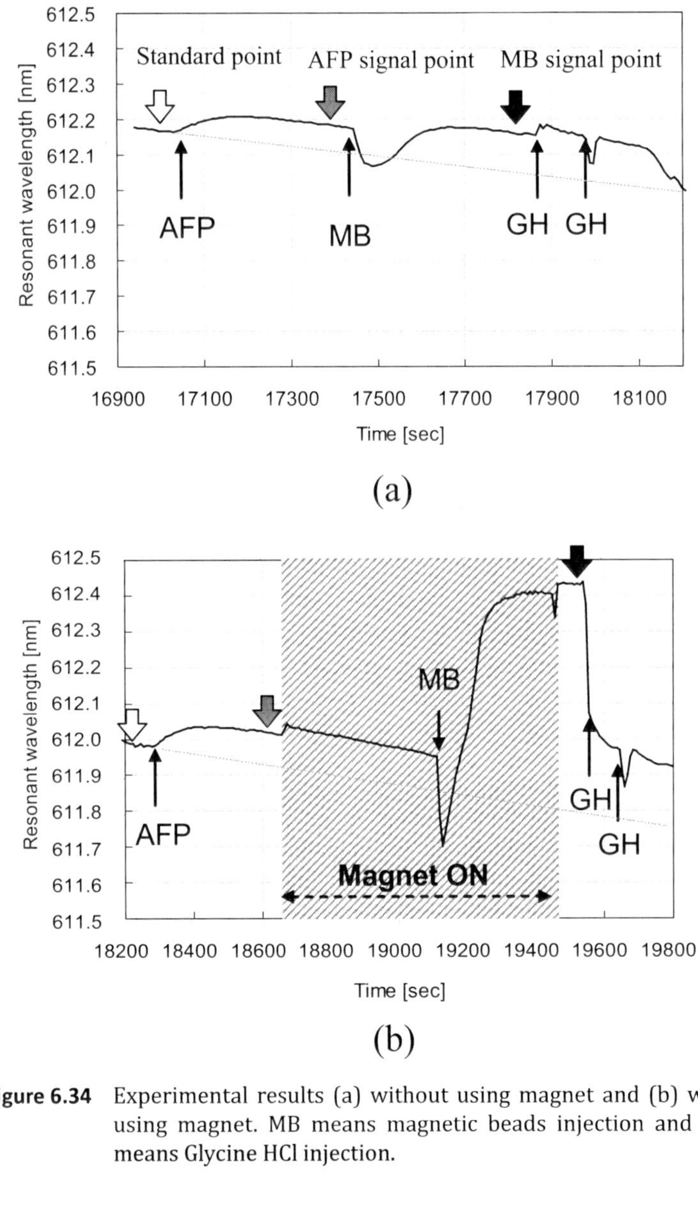

Figure 6.34 Experimental results (a) without using magnet and (b) with using magnet. MB means magnetic beads injection and GH means Glycine HCl injection.

the result of (a), the magnet was constantly off. By cancelling the drift effect due to the instability of surface immobilization, the signal shift of 0.072 nm was observed by AFP. Then the magnetic beads solution was sent over the sensor surface with the magnetic force off, and a slight signal change was observed. The total signal shift in this experiment was 0.110 nm. It was testified that the signal amplification by the sandwich method using the magnetic beads was much smaller than that using the gold colloids. The first reason of it is supposed to be that the interference between SPR and ferrite composing the magnetic beads is not so large while that between SPR and gold is very large. The second reason is supposed to be that the reaction efficiency of magnetic beads with the sensor surface is smaller than that of gold colloids since the diameter of magnetic beads is larger than that of gold colloids (diameter, 40 nm) and the shape of magnetic beads is not a complete sphere. Furthermore, the concentration of the magnetic beads was not optimized in this experiment yet.

In the experiment (b), the magnetic force was set to be on while the magnetic beads are flowing over the sensor surface. The signal shift by the reaction of AFP was about 0.089 nm which is in the same order as experiment (a). And the total signal change by AFP and the magnetic beads after turning off the magnetic force was about 0.629 nm which means that the signal is amplified as much as about 5.7 times by using the magnetic beads and magnetic force. This result verified that the reaction efficiency can be much improved by applying the magnetic force with the magnetic beads.

6.3.5.3 Injection of pre-mixture solution of AFP and magnetic beads

When the concentration of the antigen is very low, enough signal change does not occur since most of the AFP antigen does not contact with the immobilized probe protein by the previous method in which the magnetic beads are injected after the reaction between AFP and anti-AFP antibody immobilized on a sensor surface. In this part, AFP and magnetic beads on which anti-AFP antibody is immobilized are mixed for 10 min in advance, and its mixture solution is injected on a sensor surface. The experimental results obtained with the magnetic force constantly off are shown in Fig. 6.35a and b, respectively. The flow rate was 40 μL/min. When the magnetic force is off, the signal change caused by the reaction of mixture solution was 0. 150 nm. And when the magnetic force is on, the signal change was 0.538 nm which is about 3.6 times larger than that with magnetic force off. These results proved that the magnetic force can attract the magnetic beads flowing over the sensor surface down to the sensor surface and that the reaction efficiency can be much improved.

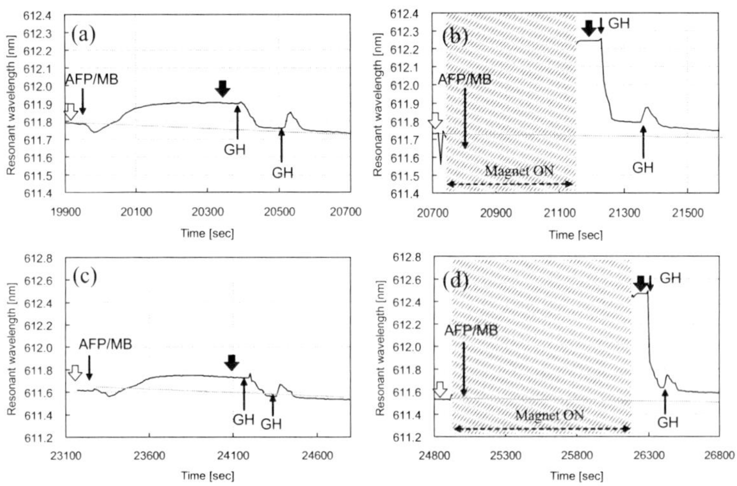

Figure 6.35 Experimental results (a), (c) without magnetic, (b), (d) with magnet. The flow rate is (a), (b) 40 μL/min and (c), (d) 20 μL/min.

As the next experiment, the effect of the flow rate is testified since the quantity of the attracted magnetic beads is affected by it. The flow rate was set at 20 μL/min here (Fig. 6.35c,d). The signal shift with the magnetic force off was 0.157 nm which is almost same as (1.05 times as large as) the result with the 40 μL/min flow rate. When the magnetic force is on, the signal change was 0.939 nm which is about 1.7 times as large as the result with the 40 μL/min flow rate and is about 6 times as large as the result with the magnetic force off. This result indicates that the signal amplification rate by the magnetic force can be apparently increased by decreasing the flow rate while the reaction rate without the magnetic force cannot significantly increase by decreasing the flow rate. This is supposed to be caused by that more magnetic beads flowing over the sensor surface is attracted to the sensor surface when the flow rate is lower. And it results in the increase of reaction rate of AFP to the probe protein on the sensor surface. Especially in the case that the concentration of AFP is very low, this method seems to be very effective to improve its sensitivity since the precious target molecules can be forced to touch with the sensor surface by the magnetic force.

Further improvement of the detection limit seems to be possible by the following optimization in the future.

(i) **Optimization of the shape of the magnetic beads.**

Since the magnetic beads used in the above experiments are not sphere but irregular shape, the antigen bound on the depressed part of the magnetic beads does not effectively bind to the probe protein on the sensor surface. To capture the precious target molecules more effectively, the magnetic beads of a sphere body is desirable.

(ii) **Optimization of the experimental parameters such as flow rate, the concentration of magnetic beads, etc.**

In the above experiment, it was proved that the sensitivity is significantly affected by the experimental parameters such as flow rate. It seems that a large potential of the sensitivity improvement remains in the optimization of these experimental parameters.

(iii) **Optimization of the application procedure of magnetic force.**

In the above experiment, the magnetic force is controlled in a very simple manner as explained below. The magnet is set "on" before the flow of magnetic beads and is set "off" after its flow. Through the experiment, it was observed that the aggregation of magnetic beads occurred on the sensor surface during on time of the magnetic force. Since the magnetic beads accumulated on its upper layer does not touch on the sensor surface, the application of magnetic force should be optimized such as switching on and off in a constant period.

(iv) **Surface modification of magnetic beads by metal layer.**

As demonstrated in the experiment based on the sandwich method, the interaction effect of the magnetic beads is smaller than that of the gold colloids. It is expected that its interaction effect can be enlarged by coating the surface of the magnetic beads with the gold layer. The amplification of the signal will be much more improved due to the attraction force of magnetic beads and the enhanced interaction effect of the gold layer.

In this subsection, it was verified that the sensitivity of SPR can be improved by a unique method which uses the magnetic force to increase the reaction efficiency of the target molecules and probe molecules on the sensor surface. Furthermore, the above ameliorations are expected to realize the detection of the target sample of a very low concentration which cannot be detected by the conventional methods.

6.4 In-line Monitoring for Purification Process

6.4.1 Basic Concept for Chromatography

Real-time detection of the small biomolecules is strongly demanded in the purification process. Chromatography is widely used in the pharmaceutical research field and for the tests of water, food, etc. Its principle is simple. The loading material is packed in the column and the dissolved sample solution is applied in it. The mixed chemicals are disassociated due to their affinity to the loading material and their sizes. At the exit of the column, the equipments such as UV detection and diagometer are generally attached to detect the purified chemicals in real time. As the problems of these conventional methods, though the presence or absence of the compound can be detected, the chemicals cannot be identified since there is no specificity in these detection tools. Therefore, the additional testing methods such as two-dimensional electrophoresis are necessary after dispensing the extracted sample to the small subjects. Furthermore, since the detection limit of the general UV absorbance is not very good (about 10 μg/mL), the target molecules with low density cannot be detected by this method.

We have proposed an in-line detection method using the nanoimprint SPR biosensor. As shown in the schematic image of the basic system (Fig. 6.36), the detection system in which the nanoimprint SPR biosensor is included is attached at the outlet of the column. The probe molecules for capturing the target molecules are prepared at the surface of sensor device. And the signal change of SPR system is monitored with the UV monitoring. By this method, the target molecules in the extracted sample solution can be detected specifically with high sensitivity in real-time. While the background effects such as the other floating materials in the column are also detected in the conventional propagating SPR, they can be eliminated in the localized SPR by the nanoimprint fabrication technology.

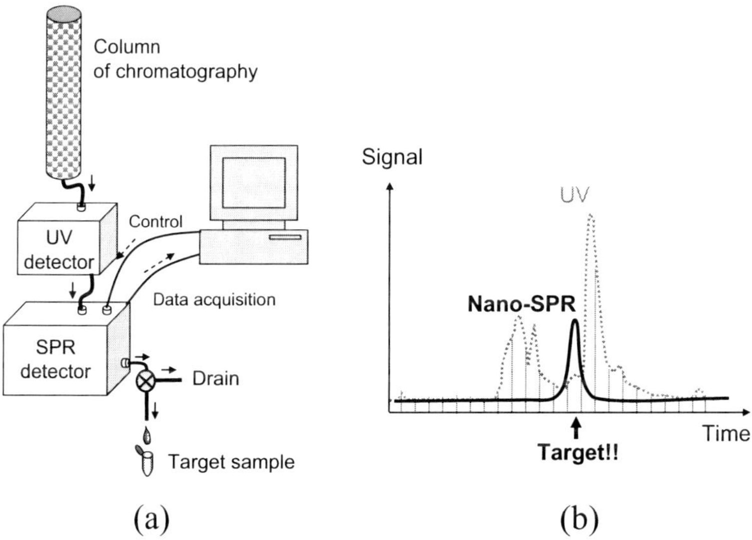

Figure 6.36 Schematics of (a) the SPR detection system for the chromatography and (b) acquisition data.

6.4.2 Detection by the Conventional Method (UV Absorption Detection)

For the comparison with the conventional method, the detection by the general UV absorption is tested.

6.4.2.1 Experimental setup

To assume the pharmaceutical process and chromatography, the experiment is conducted by using the sample, which includes the target molecules and other molecules with different concentration gradient. As the target molecule, AFP is used. And BSA (bovine serum albumin) is used as the contaminant. Fourteen separated solution samples in which these molecules are contained are prepared. The concentrations of these molecules are shown in Table 6.4. BSA (100 μg/mL) is contained in No. 10, 12. And 500 μg/mL of BSA is contained in No.11. About the AFP, 0.1 μg/mL of AFP is contained in No. 3 and 7. 0.5 μg/mL of AFP is contained in No. 4 and 6. And 3 μg/mL of AFP is contained in No. 5. As the dilution buffer, HBS-EP (10 mM HEPES-NaOH, 150 mM, NaCl 3 mM EDTA, 0.005% Tween-20, pH 7.4) is used. The concentration of AFP is 100 times smaller than that of BSA. Although the detection is conducted in real-time in the actual chromatography system, the separated sample segments are used due to the limitation of UV detection system, UV-2550 (Shimadzu Co., Japan). The calculation of absorbance is conducted at the wavelength of 280 nm.

Table 6.4 Concentration of AFP (target) and BSA (contamination) for each fraction sample

Fraction no.	AFP conc. (μg/mL)	BSA conc. (μg/mL)
1	0	0
2	0	0
3	0.1	0
4	0.5	0
5	3.0	0
6	0.5	0
7	0.1	0
8	0	0
9	0	0
10	0	100
11	0	500
12	0	100
13	0	0
14	0	0
15	0	0

6.4.2.2 Experimental result

Figure 6.37 shows the experimental result of UV absorption. Strong absorption being supposed to be caused by BSA was observed in the No. 10, 11, and 12 extracted samples. On the other hand, almost no absorption differences were observed in No. 3–7 in which AFP was contained. This seems to be that the concentration of AFP is too low to be detected by the UV absorption method. Although UV absorption detection method is widely used, it cannot realize highly sensitive and specific detection as shown in this result. Therefore, new detection technology is demanded as the in-line detection tool for the chromatography.

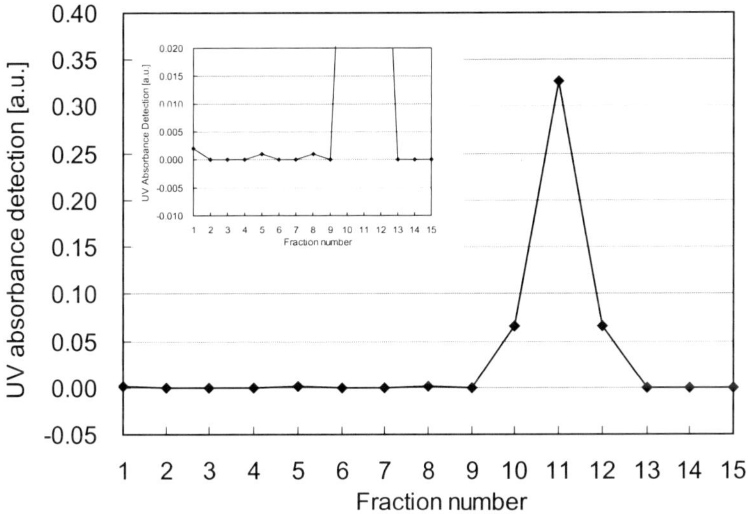

Figure 6.37 Experimental result based on UV absorbance detection.

6.4.3 Detection by Conventional SPR

In this subsection, the detection by conventional propagating SPR is testified.

6.4.3.1 Experimental setup

The experimental conditions are same as the experiment by UV absorption method. In this experiment, commercially available SPR equipment (Biacore2000, GE Healthcare) is used. On the sensor surface, anti-AFP antibody is immobilized by covalent binding after the PEG-SAM modification. In case of SPR detection, the sensor surface needs to be cleaned after the binding of the target molecules. Therefore, the surface is refreshed by using Glycine-HCl (10 mM, pH 2.0) after every detection of the separately extracted samples.

6.4.3.2 Experimental result

The experimental result is shown in Fig. 6.38. In the above UV absorption detection method, the AFP included in No. 3–7 could not be detected. In the SPR detection, however, the signal change in accordance with the concentration of the AFP was observed in No. 3–7 samples. In this result, other signal change was also detected in No. 10–12. This signal change seems to be caused by the floating BSA molecules which enter into the sensing depth of SPR so called as "bulk effect." Though an abrupt signal change occurred at No. 8, this is supposed to be the effect of the air bubble contaminant.

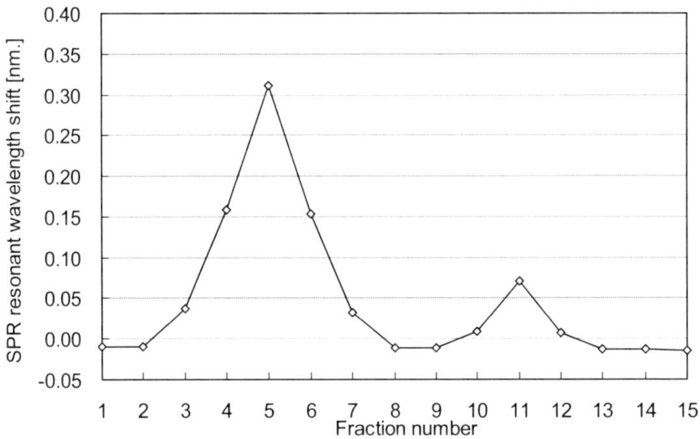

Figure 6.38 Experimental result based on SPR detection.

As shown in this experiment, low-concentration target molecules, which cannot be detected by UV absorption can be detected by using SPR. Furthermore, certain degree of specific detection can be realized by using SPR.

To eliminate the background noises (bulk effect) in this case, it is proposed that the reference sensing channel is prepared and that the signal change is compensated by using this channel. To realize high accuracy in this method, it is necessary to immobilize other antibody which has the close physical and chemical characteristic on the reference surface. Figure 6.39 shows the experimental result of the reference compensation method. As the reference probe, anti-

FLAG antibody is immobilized. Since AFP binds only on the anti-AFP antibody, the background effect can be removed by subtracting the signal change of the reference sensor on which anti-FLAG antibody is immobilized. As shown in the experimental result, the effect of background noise can be much reduced compared with the result without the compensation of the reference sensor.

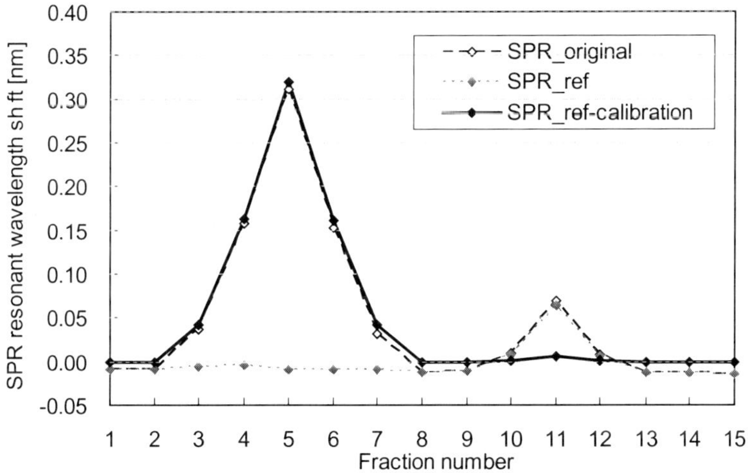

Figure 6.39 Experimental result based on SPR detection with the calibration of reference cell.

6.4.4 Detection by Nanoimprint LSPR sensor

As demonstrated in the previous result, it is proved that the sample of low concentration, which cannot be detected by UV absorption can be detected by the conventional propagating SPR. To improve its specific property, however, the compensation technique using the reference cell is necessary. This may lead to the large size and high cost of the equipment. On the other hand, in the localized SPR prepared by the nanoimprint technique, the sensing range is much smaller than that of the propagating SPR. And it is expected that the effect by the floating contaminant molecules can be reduced. To apply this technique to the in-line detection system, the same demonstration is introduced in this subsection.

6.4.4.1 Experimental setup

The experiment is conducted with the sensor device fabricated by the nanoimprint technology based on the photopolymerization method. The nanopattern is a one-dimensional groove structure whose period, width, and depth are 300 nm, about 120 nm, and about 60 nm, respectively. The thickness of the sputtered gold layer is 100 nm. By using this sensor device, the same experiment as the conventional SPR was carried out.

6.4.4.2 Experimental result

The raw experimental data and the summary of the signal change are shown in Figs. 6.40 and 6.41, respectively. Though drift was observed in the entire signal, significant signal changes depending on the concentration of AFP were observed. Almost no signal change was observed in the sample in which BSA is contained. This result verified that the background noises can be eliminated without the reference compensation by using localized SPR. This means that a small-sized, low-cost, and easy to use in-line detection system can be realized by using this technology.

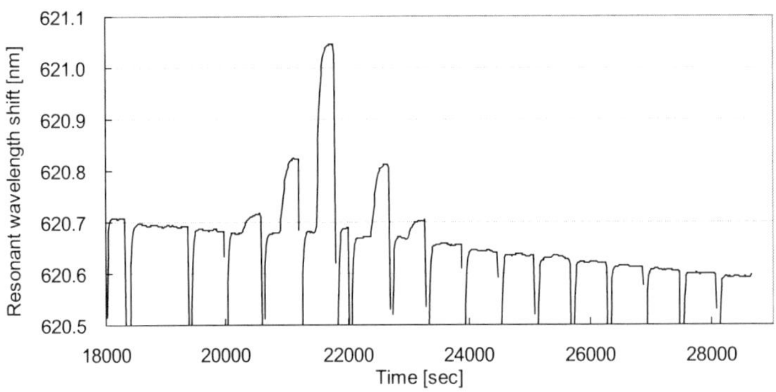

Figure 6.40 Raw sensorgram data of LSPR detection.

6.4.4.3 Surface regeneration by sheath-flow detergent

In using SPR and LSPR as the in-line detection system, it is necessary to regenerate the sensor surface to prevent the saturation of target binding. As the regeneration method, the use of detergent is the most popular one. Since the detergent leads the protein and biomolecules to the metamorphism, it is demanded that the contamination of detergent in the sample solution should be as small as possible. In the previous experiments, a large amount of the detergent is injected to fill it in the inside of the microchannel. To reduce it, we have proposed that the injection port of the detergent is prepared in the vicinity of the sensing area on the sensor surface. The regeneration of the sensor is conducted by using the sheath flow of the injected detergent. Sheath flow is the phenomenon that the flow acts as a layer and does not diffuse largely in the microchannel. In the conventional SPR, since an optical prism is attached on the backside of the sensor device, it is difficult to prepare the injection port in the vicinity of the sensing area.

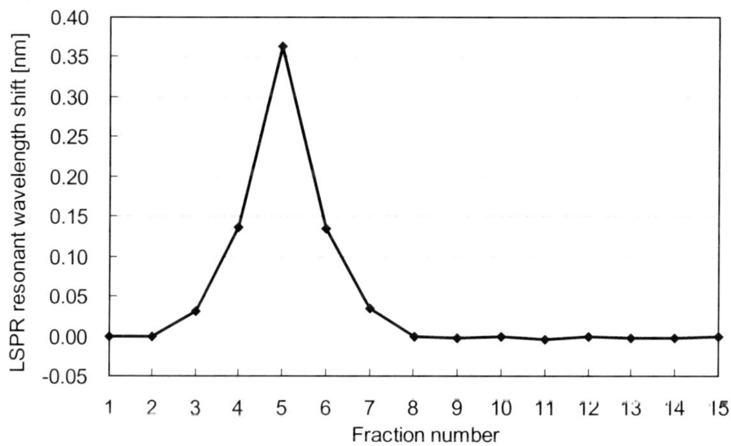

Figure 6.41 Experimental result based on LSPR detection.

To demonstrate the effect of the sheath flow, the experiment using a purified water and ethanol which are the substitutes of the buffer solution and the detergent was conducted. The experimental setup is shown in Fig. 6.42. On the sensor device, a penetration hole of 1.5 mm diameter is prepared at about 2.0 mm upstream to the sensing spot. In the main microchannel, a purified water is flowing. And the ethanol is injected from the backside of the sensor device through the penetration hole. The flow rate of both solutions was 40 μL/min. The injection volume of the ethanol is changed and the signal shift was observed with comparison to the conventional injection method. The width and height of the microchannel is

2.0 mm and 25 μm, respectively. The experimental result is shown in Fig. 6.43. In the conventional injection method, the signal change was reduced due to the diffusion of the injected ethanol when the injection volume gets smaller than 20 μL. On the other hand, in the new injection method, almost identical signal change to 50 μL injection was observed even when the injection volume was as small as 5 μL. This result verified that the sheath flow is formed when the injection port is prepared at the surface of the sensor device. And this sheath flow prevents the diffusion of the injected solution and realizes the stable reaction with much smaller injection volume. By using this technique, it is possible to reduce the contamination of sample solution by the injected detergent.

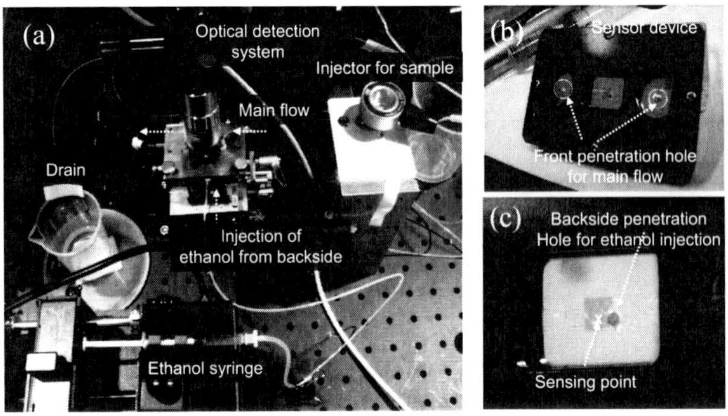

Figure 6.42 Experimental setup for sheath flow dissociation test: (a) overview of system, (b) picture of sensor device, and (c) sensing area including the backside penetration hole on the sensor device.

6.4.5 Future Applications

As described above, highly sensitive and specific real-time detection, which is difficult for the conventional detection method, can be realized by using localized SPR. By using these characteristics, this technique can be applied to the various applications. Some examples of them were introduced.

There exist the needs to detect the concentration of the specific target molecule constantly in the mass production process of the pharmaceutical product, food, etc.

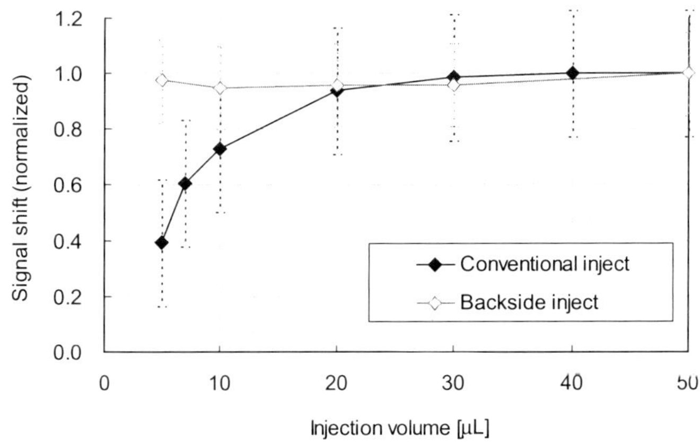

Figure 6.43 Experimental result to evaluate the effect of sheath flow.

Figure 6.44 shows the image of the nanoimprint LSPR sensor attached in the process line. This system can detect the concentration of the target molecules in real time constantly. When a certain amount of reaction occurs on the sensor surface, the detergent solution is injected to refresh the surface. The contamination of the detergent is prevented by switching the valve prepared at the downstream of the sensor device. Figure 6.45 shows simpler image to eliminate the detergent completely. The sensor line is separated from the main process line in this system. In this system, the background noise can be completely eliminated by switching the buffer solution from the contaminated sample to pure buffer sample solution. Figure 6.46 shows the system image in which the conditions of the liquid sample are controlled constantly. This system is expected to be applied to the detection of *Escherichia coli* bacteria and so on.

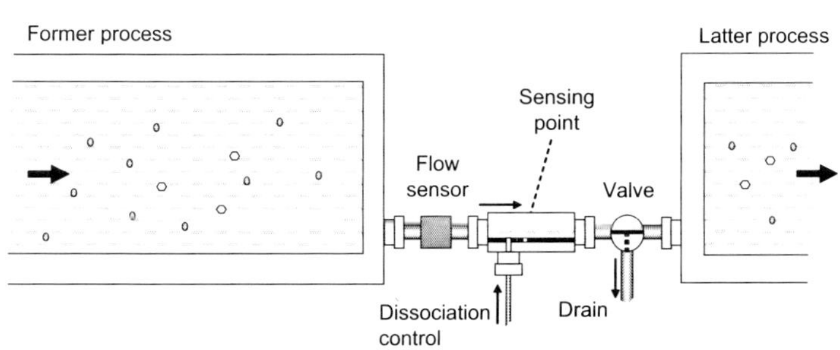

Figure 6.44 Schematic of an example of in-line sensing using nanoimprint biosensor.

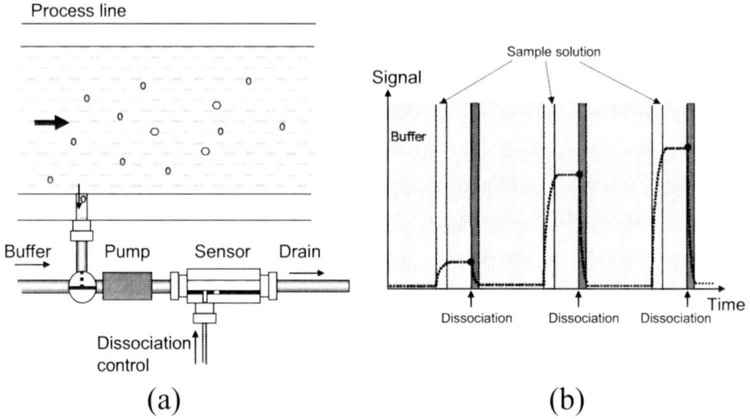

Figure 6.45 In-line application of nanoimprint biosensor: (a) schematic of system and (b) its sensorgram image.

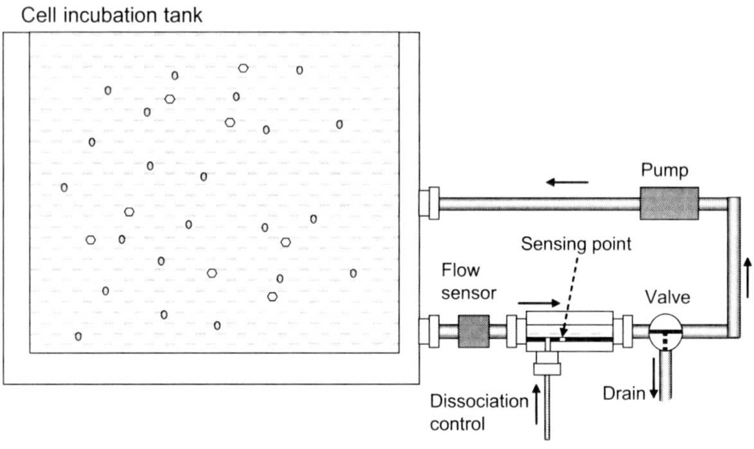

Figure 6.46 In-line application example for the quality control of the cell incubation tank.

As introduced in this section, the nanoimprint biosensor has the potential to be applied to the wide area such as laboratory tools and process equipments and so on.

Bibliography

Abrams, G. A.; Goodman, S. L.; Nealey, P. F.; Franco, M. and Murphy, C. J. (2000). Nanoscale topography of the basement membrane underlying the corneal epithelium of the rhesus macaque, *Cell Tissue Res.*, 299, pp. 39–46.

Amit, A. G.; Mariuzza, R. A.; Phillips, S. E. and Poljak, R. J. (1986). Three-dimensional structure of an antigen-antibody complex at 2.8 Å resolution. *Science*, 233, pp. 747–753.

Aoyama, S.; Kurahashi, D.; Shinohara, M. and Yamashita, T. (1996). Giant microoptics: Wide applications in liquid crystal display (LCD). In *Systems, Diffractive Optics and Micro Optics* OSA Technical Digest Series, 5, pp. 266–269.

Aoyama, S. and Yamashita, T. (1997). Planar microlens array using stumping replication method, *SPIE 3010*, pp. 11–17.

Atay, T.; Song, J. H. and Nurmikko, A. V. (2004). Strongly interacting plasmon nanoparticle pairs from dipole-dipole interaction to conductively coupled regime, *Nano Lett.*, 4(9), pp. 1627–1631.

Athey, D.; Shah, D. S. H.; Phillips, S. R. and Lakey, J. H. (2005). A manufacturable surface-biology platform for nanoapplications; cell culture, analyte detection, diagnostics sensors, *Ind. Biotechnol.*, 1, pp. 185–189.

Baek, W. H.; Seoa, I.; Yoon, T. S.; Lee, H. H.; Yun, C. M. and Kim, Y. S. (2009). Hybrid inverted bulk heterojunction solar cells with nanoimprinted TiO_2 nanopores, *Solar Energy Mater. Solar Cells*, 93, pp. 1587–1591.

Barnes, W. L.; Dereux, A. and Ebbesen, T. W. (2003). Surface plasmon subwavelength optics, *Nature*, 424, pp. 824–830.

Bernhard, C. G. (1967). Structural and functional adaptation in a visual system, *Endeavour*, 26, pp. 79–84.

Besselink, G. A. J.; Kooyman, R. P. H.; van Os, P. J. H. J.; Engbers, G. H. M. and Schasfoort, R. B. M. (2004). Signal amplification on planar and gel-type sensor surfaces in surface plasmon resonance-based detection of prostate-specific antigen, *Anal. Biochem.*, 333, pp. 165–173.

Bonroy, K.; Frederix, F.; Reekmans, G.; Dewolf, E.; Palma, R. D.; Borghs, G.; Declerck, P. and Goddeeris, B. (2006). Comparison of random and oriented immobilisation of antibody fragments on mixed self-assembled monolayers, *J. Immunol. Methods*, 312, pp. 167–181.

Born, M. and Wolf, E. (1975). *Principles of optics* (5th edition), Pergamon Press, Oxford.

Byun, K. M.; Kim, S. J. and Kim, D. (2006). Profile effect on the feasibility of extinction-based localized surface plasmon resonance biosensors with metallic nanowires, *Appl. Opt.*, 45(14): pp. 3382–3389.

Chambers, D. M. and Nordin, G. P. (1999). Stratified volume diffractive optical elements as high-efficiency gratings, *J. Opt. Soc. Am. A*, 16(5), pp. 1184–1193.

Chilwell, J. and Hodgkinson, I. (1984). Thin-films field-transfer matrix theory of planar mutilayer wave guides and reflection from prism-loaded waveguides, *J. Opt. Soc. Am.*, A1(7), pp. 742–753.

Cho, I.-H.; Paek, E.-H.; Lee, H.; Kang, J. Y.; Kim, T. S. and Paek, S. H. (2007). Site-directed biotinylation of antibodies for controlled immobilization on solid surfaces, *Anal. Biochem.*, 365, pp. 14–23.

Choi, J.; Nordquist, K; Cherala, A; Casoose, L.; Gehoski, K.; Dauksher, W. J.; Sreenivasan, S. V. and Resnick, D. J. (2005). Distortion and overlay performance of UV step and repeat imprint lithography, *Microelectron. Eng.*, 78–79(1–4), pp. 633–640.

Choi, Y. J.; Sreenivasan, S. V. and Choi, B. J. (2008). Kinematic design of large displacement precision XY positioning stage by using cross strip flexure joints and over-constrained mechanism, *Mech. Mach. Theory*, 43(6), pp. 724–737.

Chou, S. Y.; Keimel, C. and Gu, J. (2002). Ultrafast and direct imprint of nanostructures in silicon, *Nature*, 417, pp. 835–837.

Chou, S. Y. and Krauss, P. R. (1997). Imprint lithography with sub-10 nm feature size and high throughput, *Microelectron. Engineer.*, 35, pp. 237–240.

Chou, S. Y.; Krauss, P. R. and Renstrom, P. J. (1995). Imprint of sub-25 nm vias and trenches in polymers, *Appl. Phys. Lett.*, 67, pp. 3114–3116.

Chou, S. Y.; Zhuang, L. and Sun, X. (1998). Multilayer resist methods for nanoimprint lithography on nonflat surfaces, *J. Vac. Sci. Technol. B.*, 16, pp. 3922–3925.

Chung, J. W.; Park, J. M.; Bernhardt, R. and Pyun, J. C. (2006). Immunosensor with a controlled orientation of antibodies by using neutravidin-protein a complex at immunoaffinity layer, *J. Biotechnol.*, 126, pp. 325–333.

Colburn, M.; Bailey, T.; Choi, B. J.; Ekerdt, J. G.; Sreenivasan, S. V. and Willson, C. G. (2001). Development and advantages of step and flash imprint lithography, *Solid State Technol.*, 46(7), pp. 67–78.

Colburn, M.; Johnson, S.; Stewart, M.; Damle, S.; Bailey, T.; Choi, B.; Wedlake, M.; Michaelson, T.; Sreenivasan, S. V.; Ekerdt, J. and Willson, C. G. (1999). Step and flash imprint lithography: A new approach to high-resolution patterning, *Proc. SPIE—Emerg. Lith. Tech.*, 3676, pp. 379–389.

Crosson, C.; Thomas, D. and Rossi, C. (2010). Quantification of immunoglobulin G in bovine and caprine milk using a surface plasmon resonance-based immunosensor, *J. Agric. Food Chem.*, 58, pp. 3259–3264.

Cullen, D. C.; Brown, R. G. and Lowe, C. R. (1987). Detection of immuno-complex formation via surface plasmon resonance on gold-coated diffraction gratings, *Biosensors*, 3(4), pp. 211–225.

Curtis, A. and Wilkinson, C. (1997). Topographical control of cells, *Biomaterials*, 18, pp. 1573–1583.

Deguchi, K.; Takeuchi, N. and Shimizu, A. (2002). Evaluation of pressure uniformity using a pressure-sensitive film and calculation of wafer distortions caused by mold press in imprint lithography, *Jpn. J. Appl. Phys.*, 41, pp. 4178–4181 (in Japanese).

Ducre, J. J. and Zengerle, R. (2012). *Microfluidics (Microtechnology and MEMS)*, Springer-Verlag: Berlin and Heidelberg.

Duval Malinsky, M.; Kelly, K. L.; Schatz, G. C. and Van Duyne, R. P. (2001). Nanosphere lithography: Effect of substrate on the localized surface plasmon resonance spectrum of silver nanoparticles, *J. Phys. Chem. B.*, 105(12), pp. 2343–2350.

Ebbesen, T. W.; Lezec, H. J.; Ghaemi, H. F.; Thio, T. and Wolff, P. A. (1998). Extraordinary optical transmission through sub-wavelength hole arrays, *Nature*, 391, pp. 667–669.

Faraday, M. (1857). The Bakerian lecture. Experimental relations of gold (and other metals) to light, *Philos. Trans.*, 147, pp. 145–181.

Fröhlich, H. (1949). *Theory of Dielectrics*, Oxford University Press, London.

Fujikata, J.; Ishi, T.; Makita, K.; Baba, T. and Ohashi, K. (2005). Highly enhanced speed and efficiency of Si nano-photodiode with a surface-plasmon antenna, *Int. Conf. SSDM*, E-3-3.

Gobi, K. V.; Iwasaka, H. and Miura, N. (2007). Self-assembled PEG monolayer based SPR immunosensor for label-free detection of insulin, *Biosens. Bioelectron.*, 22, pp. 1382–1389.

Grann, E. B.; Moharam, M. G. and Pommet, D. A. (1995). Optimal design for antireflective tapered two-dimensional subwavelength grating structures, *J. Opt. Soc. Am.*, 12(2), pp. 333–339.

Grigalinas, V.; Tamuleviius, S.; Tomainas, R.; Kopustinskas, V.; Guobien, A. and Jucius, D. (2004). Laser pulse assisted nanoimprint lithography, *Thin Solid Films*, 453–454, pp. 13–15.

Guo, L.; Krauss, P. R. and Chou S. Y. (1997). Nanoscale silicon field effect transistors fabricated using imprint lithography, *Appl. Phys. Lett.*, 71(13), pp. 1881–1883.

Guo, L.; Leobandung, E.; Zhuang, L. and Chou, S. Y. (1997). Fabrication and characterization of room temperature silicon single electron memory, *J. Vac. Sci. Technol. B.*, 15, pp. 2840–2843.

Ha, T. H.; Jung, S. O.; Lee, J. M.; Lee, K. Y.; Lee, Y.; Park, J. S. and Chung, B. H. (2007). Oriented immobilization of antibodies with GST-fused multiple Fc-specific B-domains on a gold surface, *Anal. Chem.*, 79, pp. 546–556.

Haes, A. J.; Hall, W. P.; Chang, L.; Klein, W. L. and Van Duyne, R. P. (2004). A localized surface plasmon resonance biosensor: First steps toward an assay for Alzheimer's disease, *Nano Lett.*, 4(6), pp. 1029–1034.

Haes, A. J. and Van Duyne, R. P. (2002). A nanoscale optical biosensor: Sensitivity and selectivity of an approach based on the localized surface plasmon resonance spectroscopy of triangular silver nanoparticles, *J. Am. Chem. Soc.*, 124 (35), pp. 10596–10604.

Hamada, H.; Fujii, A.; Mizuguchi, Y.; Shibatani, T.; Funada, F. and Awane, K. (1995). A new high definition microlens array built in p-Si TFT-LCD panel, *Proc. Asia Display'95*, pp. 887–890.

Hayashi, S. (2002). *Nano photonics handbook*, p. 86 (in Japanese), Asakura Press: Tokyo, Japan.

Heidari, B.; Maximov, I. and Montelius, L. (2000). Nanoimprint lithography at the 6 in. wafer scale, *J. Vac. Sci. Technol. B.*, 18(6), pp. 3557–3560.

Her, H. J.; Kim, J. M.; Kang, C. J. and Kim, Y. S. (2008). Hybrid photovoltaic cell with well-ordered nanoporous titania–P3HT by nanoimprinting lithography, *J. Phys. Chem. Solids.*, 69, pp. 1301–1304.

Hoff, A. D.; Cheng, L. J.; Meyhöfer, E.; Guo, L. J. and Hunt A. J. (2004). Nanoscale protein patterning by imprint lithography, *Nano Lett.*, 4(5), pp. 853–857.

Homola, J.; Koudela, I. and Yee S. S. (1999). Surface plasmon resonance sensors based on diffraction gratings and prism couplers: Sensitivity comparison, *Sens. Actuators B*, 54, pp. 16–24.

Hong, E. J.; Byeon, K. J.; Park, H.; Hwang, J.; Lee, H.; Choi, K. and Jung, G. Y. (2009). Fabrication of moth-eye structure on p-GaN layer of GaN-based LEDs for improvement of light extraction, *Mater. Sci. Eng. B*, 163(3), pp. 170–173.

Hu, W.; Yim, E. K. F.; Reano, R. M.; Leong, K. W. and Pang, S. W. (2005). Effects of nanoimprinted patterns in tissue-culture polystyrene on cell behavior, *J. Vac. Sci. Technol. B*, 23(6), pp. 2984–2989.

Huang, L.; Reekmans, G.; Saerens, D.; Friedt, J.-M.; Frederix, F.; Francis, L.; Muyldermans, S.; Campitelli, A. and Hoof, C. V. (2005). Prostate-specific antigen immunosensing based on mixed self-assembled monolayers, camel antibodies, and colloidal gold enhanced sandwich assays, *Biosens. Bioelectron.*, 21, pp. 483–490.

Huang, N.-P.; Vörös, J.; De Paul, S. M.; Textor, M. and Spencer, N. D. (2002). Biotin-derivatized poly(L-lysine)-g-poly(ethylene glycol): A novel polymeric interface for bioaffinity sensing, *Langmuir*, 18, pp. 220–230.

Ishi, T.; Fujikata, J.; Makita, K.; Baba, T. and Ohashi, K. (2005). Si nano-photodiode with a surface plasmon antenna, *Jpn. J. Appl. Phys.*, 44, pp. L364–L366 (in Japanese).

Ishizuka-Katsura, Y.; Wazawa, T.; Ban, T.; K. Morigaki, K. and Aoyama, S. (2008). Biotin-containing phospholipid vesicle layer formed on self-assembled monolayer of a saccharide-terminated alkyl disulfide for surface plasmon resonance biosensing, *J. Biosci. Bioeng.*, 105, pp. 527–535.

Jans, K.; Bonroy, K.; Palma, R. D.; Reekmans, G.; Jans, H.; Laureyn, W.; Smet, M.; Borghs, G. and Maes, G. (2008). Stability of mixed PEO-thiol SAMs for biosensing applications, *Langmuir*, 24, pp. 3949–3954.

Jarem, J. M. (2002). Validation and numerical convergence of the Hankel-Bessel and Mathieu rigorous coupled wave analysis algorithms for radially and azimuthally inhomogeneous, elliptical, cylindrical systems, *Prog. Electromag. Res.*, 36, pp. 153–177.

Jarem, J. M. and Banerjee, P. P. (1998). Rigorous coupled-wave analysis of photorefractive reflection gratings, *J. Opt. Soc. Am. B*, 15(7), pp. 2099–2106.

Jensen, T. R.; Malinsky, M. D.; Haynes, C. L. and Van Duyne, R. P. (2000). Nanosphere lithography: Tunable localized surface plasmon resonance spectra of silver nanoparticles, *J. Phys. Chem. B.*, 104(45), pp. 10549–10556.

Johnsson, B.; Löfås, S. and Lindquist, G. (1991). Immobilization of proteins to a carboxymethyldextran-modified gold surface for biospecific interaction analysis in surface plasmon resonance sensors, *Anal. Biochem.*, 198, pp. 268–277.

Jönsson, U.; Fägerstam, L.; Ivarsson, B.; Johnsson, B.; Karlsson, R.; Lundh, K.; Löfås, S.; Persson, B.; Roos, H.; Rönnberg, I.; Sjölander, S.; Stenberg, E.; Ståhlberg, R.; Urbaniczky, C.; Östlin, H. and Malmqvist, M. (1991). Real-time biospecific interaction analysis using surface plasmon resonance and a sensor chip technology, *Biotechniques*, 11, pp. 620–627.

Jung, L. S.; Campbell, C. T.; Chinowsky, T. M.; Mar, M. N. and Yee, S. S. (1998). Quantitative interpretation of the response of surface plasmon resonance sensors to adsorbed films, *Langmuir*, 14(19), pp. 5636–5648.

Jung, Y.; Lee, J. M.; Jung, H. and Chung, B. H. (2007). Self-directed and self-oriented immobilization of antibody by protein G-DNA conjugate, *Anal. Chem.*, 79, pp. 6534–6541.

Karlsson, R.; Michaelsson, A. and Mattsson, L. (1991). Kinetic analysis of monoclonal antibody-antigen interactions with a new biosensor based analytical system, *J. Immunol. Methods*, 145, pp. 229–240.

Katz, B. A.; Cass, R. T.; Liu, B.; Arze, R. and Collins, N. (1995). Topochemical catalysis achieved by structure-based ligand design, *J. Biol. Chem.*, 270, pp. 31210–31218.

Kausaite-Minkstimiene, A.; Ramanaviciene, A.; Kirlyte, J. and Ramanavicius, A. (2010). Comparative study of random and oriented antibody immobilization techniques on the binding capacity of immunosensor, *Anal. Chem.*, 82, pp. 6401–6408.

Kawaguchi, T.; Shankaran, D. R.; Kim, S. J.; Gobi, K. V.; Matsumoto, K.; Toko, K. and Miura, N. (2007). Fabrication of a novel immunosensor using functionalized self-assembled monolayer for trace level detection of TNT by surface plasmon resonance, *Talanta*, 72, pp. 554–560.

Kelley, S. C.; Deluga, G. A. and Smyrl, W. H. (2002). Miniature fuel cells fabricated on silicon substrates, *AIChE J.*, 48(5), pp. 1071–1082.

Kim, F.; Song, J. H. and Yang, P. (2002). Photochemical synthesis of gold nanorods, *J. Am. Chem. Soc.*, 124(48), pp. 14316–14317.

Kim, K.; Jeong, J.; Park, S.; Choi, D.; Choi, J. and Lee, E. (2009). Development of a very large-area ultraviolet imprint lithography process, *Microelectron. Eng.*, 86(10), pp. 1983–1988.

Kim, S.; Jung, J. M.; Choi, D. G.; Jung, H. T. and Yang, S. M. (2006). Patterned arrays of au rings for localized surface plasmon resonance, *Langmuir,* 22(17), pp. 7109–7112.

Kim, S. J.; Gobi, K. V.; Iwasaka, H.; Tanaka, H. and Miura, N. (2007). Novel miniature SPR immunosensor equipped with all-in-one multi-microchannel sensor chip for detecting low-molecular-weight analytes, *Biosens. Bioelectron.,* 23, pp. 701–707.

Kothare, M. V. (2006). Dynamics and control of integrated microchemical systems with application to micro-scale fuel processing, *Comp. Chem. Eng.*, 30, pp. 1725–1734.

Krauss, P. R. and Chou, S. Y. (1995). Fabrication of planar quantum magnetic disk structure using electron beam lithography, reactive ion etching, and chemical mechanical polishing, *J. Vac. Sci. Technol. B.*, 13(6), pp. 2850–2852.

Kretschmann, E. and Raether, H. (1968). Radiative decay of non radiative surface plasmons excited by light, *Z. Naturf.,* A23, pp. 2135–2136.

Ku, J. and Stroeve, P. (2004). Protein diffusion in charged nanotubes: "On-off" behavior of molecular transport, *Langmuir,* 20, pp. 2030–2032.

Kumar, A. and Whitesides, G. M. (1993). Features of gold having micrometer to centimeter dimensions can be formed through a combination of stamping with an elastomeric stamps and an alkanethiol "ink" followed by chemical etching, *Appl. Phys. Lett.*, 63, pp. 2002–2004.

Kume, T.; Hayashi, S.; Ohkuma, H. and Yamamoto, K. (1993). Enhancement of photoelectric conversion efficiency in copper phthalocyanine solar cell by surface plasmon excitation, *Jpn. J. Appl. Phys.*, 32(8), pp. 3486–3492 (in Japanese).

Kume, T.; Hayashi, S.; Ohkuma, H. and Yamamoto, K. (1995). Enhancement of photoelectric conversion efficiency in copper phthalocyanine solar cell: White light excitation of surface plasmon polaritons, *Jpn. J. Appl. Phys.*, 34(12A), pp. 6448–6451 (in Japanese).

Kwon, Y.; Han, Z.; Karatan, E.; Mrksich, M. and Kay, B. K. (2004). Antibody arrays prepared by cutinase-mediated immobilization on self-assembled monolayers, *Anal. Chem.*, 76, pp. 5713–5720.

Kyo, M.; Usui-Aoki, K. and Koga, H. (2005). Label-free detection of proteins in crude cell lysate with antibody arrays by a surface plasmon resonance imaging technique, *Anal. Chem.*, 77, pp. 7115–7121.

Lahiri, J.; Isaacs, L.; Tien, J. and Whitesides, G. M. (1999). A strategy for the generation of surfaces presenting ligands for studies of binding based on active aster as a common reactive intermediate: A surface plasmon resonance study, *Anal. Chem.*, 71, pp. 777–790.

Lazzarino, F.; Gourgon, C.; Schiavone, P. and Perret, C. (2004). Mold deformation in nanoimprint lithography, *J. Vac. Sci. Technol. B.*, 22(6), pp. 3318–3322.

Le Brun, A. P.; Holt, S. A.; Shah, D. S. H.; Majkrzak, C. F. and Lakey, J. H. (2011). The structural orientation of antibody layers bound to engineered biosensor surfaces. *Biomaterials*, 32, pp. 3303–3311.

Lee, E. S.; Jeong, J. H.; Sim, Y. S.; Kim, K. D.; Choi, D. G. and Choi, J. H. (2006). High-throughput step-and-repeat UV-nanoimprint lithography, *Curr. Appl. Phys.*, 6, pp. e92–e98.

Lee, J. M.; Park, H. K.; Jung, Y.; Kim, J. K.; Jung, S. O. and Chung, B. H. (2007). Direct immobilization of protein G variants with various numbers of cysteine residues on a gold surface, *Anal. Chem.*, 79, pp. 2680–2687.

Lehnert, D.; Wehrle-Haller, B.; David, C.; Weiland, U.; Ballestrem, C.; Imhof, B. A. and Bastmeyer, M. (2004). Cell behaviour on micropatterned substrata: Limits of extracellular matrix geometry for spreading and adhesion, *J. Cell Sci.*, 117, pp. 41–52.

Lezec, H. J.; Degiron, A.; Devaux, E.; Linke, R. A.; Martin-Moreno, L.; Garcia-Vidal, F. J. and Ebbesen, T. W. (2002). Beaming light from a subwavelength aperture, *Science*, 297, pp. 820–822.

Liedberg, B.; Nylander, C. and Lunstrom, I (1983). Surface plasmon resonance for gas detection and biosensing, *Sens. Actuators*, 4, pp. 299–304.

Lindquist, N. C.; Luhman, W. A.; Oh, S. H. and Holmes, R. J. (2008). Plasmonic nanocavity arrays for enhanced efficiency in organic photovoltaic cells, *Appl. Phys. Lett.*, 93(12), p. 123308.

Lodder, J. C. (2004). Methods for preparing patterned media for high-density recording, *J. Magn. Magn. Mater.*, 272–276, pp. 1692–1697.

Löfås, S.; Johnsson, B.; Edström, Å.; Hansson, A.; Lindquist, G.; Hillgren, R.-M. M. and Stigh, L. (1995). Methods for site controlled coupling to carboxymethyldextran surfaces in surface plasmon resonance sensors. *Biosens. Bioelectron.*, 10, pp. 813–822.

Löfås, S. and Mcwhirter, A. (2006). The art of immobilization for SPR sensors, *Springer Ser. Chem. Sens. Biosens.*, 4, pp. 117–151.

Lu, G. Q.; Wang, C. Y.; Yen, T. J. and Zhang, X. (2004). Development and characterization of a silicon-based micro direct methanol fuel cell, *Electrochim. Acta*, 49, pp. 821–828.

Masson, J. -F.; Battaglia, T. M.; Khairallah, P.; Beaudoin, S. and Booksh, K. S. (2007). Quantitative measurement of cardiac markers in undiluted serum, *Anal. Chem.*, 79, pp. 612–619.

McClelland, G. M.; Hart, M. W.; Rettner, C. T.; Best, M. E.; Carter, K. R. and Terris, B. D. (2002). Nanoscale patterning of magnetic islands by imprint lithography using a flexible mold, *Appl. Phys. Lett.*, 81, pp. 1483–1485.

Meyers, J. P.; and Maynard, H. L. (2002). Design considerations for miniaturized PEM fuel cells, *J. Power Sources*, 109(1), pp. 76–88.

Moore, G. E. (1965). Cramming more components onto integrated circuits, *Electronics*, 38, pp. 114–117.

Murphy, C. J. and Jana, N. R. (2002). Controlling the aspect ratio of inorganic nanorods and nanowires, *Adv. Mater.*, 14, pp. 80–82.

Nagatomo, K.; Kawaguchi, T.; Miura, N.; Toko, K. and Matsumoto, K. (2009). Development of a sensitive surface plasmon resonance immunosensor for detection of 2,4-dinitrotoluene with a novel oligo (ethylene glycol)-based sensor surface, *Talanta*, 79, pp. 1142–1148.

Nath, N. and Chilkoti, A. (2004). Label-free biosensing by surface plasmon resonance of nanoparticles on glass: Optimization of nanoparticle size, *Anal. Chem.*, 76, pp. 5370–5378.

Nilsson, C. E.; Abbas, S.; Bennemo, M.; Larsson, A.; Hämäläinen, M. D. and Frostell-Karlsson, Å. (2010). A novel assay for influenza virus quantification using surface plasmon resonance, *Vaccine*, 28, pp. 759–766.

Nishizaki, O.; Fujioka, R. and Ito, Y. (2007). Development of D-ARS (Dustproof Anti-Reflection Structure) by using nanoimprint technology, *Omron Tech.*, 48(1), pp. 29–34.

Nylander, C.; Liedberg, B. and Lind, T. (1982/1983). Gas detection by means of surface plasmon resonance, *Sens. Actuators*, 3, pp. 79–88.

Odom, T. W.; Love, J. C.; Wolfe, D. B.; Paul, K. E. and Whitesides, G. M. (2002). Improved pattern transfer in soft lithography using composite stamps, *Langmuir*, 18, pp. 5314–5320.

Odom, T. W.; Thalladi, V. R.; Love, J. C. and Whitesides, G. M. (2002). Generation of 30–50 nm structures using easily fabricated, composite PDMS masks, *J. Am. Chem. Soc.*, 124, pp. 12112–12113.

O'Hayre, R.; Fabian, T.; Lee, S. J. and Prinz, F. B. (2003). Lateral ionic conduction in planar array fuel cells, *J. Electrochem. Soc.*, 150(4), pp. A430–A438.

Okamoto, T.; Yamaguchi, I. and Kobayashi, T. (2000). Local plasmon sensor with gold colloid monolayers deposited upon glass substrates, *Opt Lett.*, 25(6), pp. 372–374.

Otto, A. (1968). Excitation of non-radiative surface plasma waves in silver by the method of frustrated total reflection, *Z. Phys.*, 216, pp. 398–410.

Otto, M.; Bender, M.; Hadam, B.; Spangenberg, B. and Kurz, H. (2000). Characterization and application of a UV-based imprint technique, *Microelectron. Eng.*, 57/58, pp. 361–366.

Park, T. J.; Hyun, M. S.; Lee, H. J.; Lee, S. Y. and Ko, S. (2009). A self-assembled fusion protein-based surface plasmon resonance biosensor for rapid diagnosis of severe acute respiratory syndrome, *Talanta*, 79, pp. 295–301.

Pale-Grosdemange, C.; Simon, E. S.; Prime, K. L. and Whitesides, G. M. (1991). Formation of self-assembled monolayers by chemisorption of derivatives of oligo(ethylene glycol) of structure $HS(CH_2)_{11}(OCH_2CH_2)_mOH$ on gold, *J. Am. Chem. Soc.*, 113, pp. 12–20.

Pei, Z.; Anderson, H.; Myrskog, A.; Dunér, G.; Ingemarsson, B. and Aastrup, T. (2010). Optimizing immobilization on two-dimensional carboxyl surface: pH dependence of antibody orientation and antigen binding capacity, *Anal. Biochem.*, 398, pp. 161–168.

Polonschii, C.; David, S.; Tombelli, S.; Mascini, M. and Gheorghiu, M. (2010). A novel low-cost and easy to develop functionalization platform. Case study: Aptamer-based detection of thrombin by surface plasmon resonance, *Talanta*, 80, pp. 2157–2164.

Prime, K. L. and Whitesides, G. M. (1991). Self-assembled organic monolayers: Model systems for studying adsorption of proteins at surfaces, *Science*, 252, pp. 1164–1167.

Prime, K. L. and Whitesides, G. M. (1993). Adsorption of proteins onto surfaces containing end-attached oligo(ethylene oxide): A model system using self-assembled monolayers, *J. Am. Chem. Soc.*, 115, pp. 10714–10721.

Rechberger, W.; Hohenau, A.; Leitner, A.; Krenn, J. R.; Lamprecht, B. and Aussenegg, F. R. (2003). Optical properties of two interacting gold nanoparticles, *Opt. Comm.*, 220, pp. 137–141.

Ross, A.; Smith, H. I.; Savas, T.; Schattenburg, M.; Farhoud, M.; Hwang, M.; Walsh, M.; Abraham, M. C. and Ram, R. J. (1999). Fabrication of patterned media for high density magnetic storage, *J. Vac. Sci. Technol. B.*, 17, pp. 3168–3176.

Sepúlveda, B.; Angelomé, P. C.; Lechuga, L. M. and Liz-Marzán, L. M. (2009). LSPR-based nanobiosensors, *Nano Today*, 4(3), pp. 244–251.

Shah, D. S.; Thomas, M. B.; Phillips, S.; Cisneros, D. A.; Le Brun, A. P.; Holt, S. A. and Lakey, J. H. (2007). Self-assembling layers created by membrane proteins on gold, *Biochem. Soc. Trans.*, 35, pp. 522–526.

Shi, H.; Tsai, W. B.; Garrison, M. D.; Ferrari, S. and Ratner, B. D. (1999). Template-imprinted nanostructured surfaces for protein recognition, *Nature*, 398, pp. 593–597.

Sigal, G. B.; Bamdad, C.; Barberis, A.; Strominger, J. and Whitesides, G. M. (1996). A self-assembled monolayer for the binding and study of histidine-tagged proteins by surface plasmon resonance, *Anal. Chem.*, 68, pp. 490–497.

Sigal, G. B.; Mrksich, M. and Whitesides, G. M. (1998). Effect of surface wettability on the adsorption of proteins and detergents, *J. Am. Chem. Soc.*, 120, pp. 3464–3473.

Situ, C.; Wylie, A. R. G.; Douglas, A. and Elliott, C. T. (2008). Reduction of severe bovine serum associated matrix effects on carboxymethylated dextran coated biosensor surfaces, *Talanta*, 76, pp. 832–836.

Skottrup, P.; Hearty, S.; Frøkiær, H.; Leonard, P.; Hejgaard, J.; O'Kennedy, R.; Nicolaisen, M. and Justesen, A. F. (2007). Detection of fungal spores using a generic surface plasmon resonance immunoassay, *Biosens. Bioelectron.*, 22, pp. 2724–2729.

Sotomayor Torres, C. M.; Zankovych, S.; Seekamp, J.; Kam, A. P.; Clavijo Cedeno, C.; Hoffmann, T.; Ahopelto, J.; Reuther, F.; Pfeiffer, K.; Bleidiessel, G.; Gruetzner, G.; Maximov, M. V. and Heidari, B. (2003). Nanoimprint lithography: An alternative nanofabrication approach, *Mater. Sci. Eng.*, 23, pp. 23–31.

Spears, D. L. and Smith, H. I. (1972). High-resolution pattern replication using soft X rays, *Electr. Lett.*, 8, pp. 102–104.

Stapleton, S.; Bradshaw, B. and O'Kennedy, R. (2009). Development of a surface plasmon resonance-based assay for the detection of *Corynebacterium pseudotuberculosis* infection in sheep, *Anal. Chim. Acta*, 651, pp. 98–104.

Stenberg, E.; Persson, B.; Roos, H. and Urbaniczky, C. (1991). Quantitative determination of surface concentration of protein with surface plasmon resonance using radiolabeled proteins, *J. Colloid Interface Sci.*, 143, pp. 513–526.

Stephan, L. and Mostafa A. E. (1999). Spectral properties and relaxation dynamics of surface plasmon electronic oscillations in gold and silver nanodots and nanorods, *J. Phys. Chem. B*, 103, pp. 8410–8426.

Tamura, T.; Nakao, M.; Ozawa, A. and Masuda, H. (1998). Direct nano-printing on Al substrate using a SiC mold, *J. Vac. Sci. Technol. B.*, 16(3), pp. 1145–1149.

Tan, H.; Gilbertson, A. and Chou, S. Y. (1998). Roller nanoimprint lithography, *J. Vac. Sci. Technol. B.*, 166, pp. 3926–3928.

Taylor, A. D.; Lucas, B. D.; Guo, L. J. and Thompson, L. T. (2007). Nanoimprinted electrodes for micro-fuel cell applications, *J. Power Sources*, 171, pp. 218–223.

Teramura, Y. and Iwata, H. (2007). Label-free immunosensing for α-fetoprotein in human plasma using surface plasmon resonance, *Anal. Biochem.*, 365, pp. 201–207.

Terrettaz, S.; Ulrich, W. -P.; Vogel, H.; Hong, Q.; Dover, L. G. and Lakey, J. H. (2002). Stable self-assembly of a protein engineering scaffold on gold surfaces, *Protein Sci.*, 11, pp. 1917–1925.

Terris, B. D. (2009). Fabrication challenges for patterned recording media, *J. Magn. Magn. Mater.*, 321(6), pp. 512–517.

Thio, T.; Pellerin, K. M.; Linke, R. A.; Lezec, H. J. and Ebbesen, T. W. (2001). Enhanced light transmission through a single subwavelength aperture, *Opt. Lett.*, 26, pp. 1972–1974.

Torrance, L.; Ziegler, A.; Pittman, H.; Paterson, M.; Toth, R. and Eggleston, I. (2006). Oriented immobilisation of engineered single-chain antibodies to develop biosensors for virus detection, *J. Vir. Methods*, 134, pp. 164–170.

Truskett, V. N. and Watts, M. P. C. (2006). Trends in imprint lithography for biological applications, *Trends Biotechnol.*, 24(7), pp. 312–317.

Uludağ, Y. and Tothill, I. E. (2010). Development of a sensitive detection method of cancer biomarkers in human serum (75%) using a quartz crystal microbalance sensor and nanoparticles amplification system, *Talanta*, 82, pp. 277–282.

Usui, H.; Matsumoto, H.; Gendrin, R. and Nishikawa, T. (2001). Computer experiments on a three-wave in association with microwave power transmission in space plasma, *IEICE Trans.*, E84-B(9), pp. 2566–2573.

Vial, S.; Pastoriza-Santos, I.; Perez-Juste, J. and Liz-Marzan. L. M. (2007). Plasmon coupling in layer-by-layer assembled gold nanorod films, *Langmuir*, 23(8), pp. 4606–4611.

Watanabe, T.; Muranaka, N.; Iijima, I. and Hohsaka, T. (2007). Position-specific incorporation of biotinylated non-natural amino acids into a protein in a cell-free translation system, *Biochem. Biophys. Res. Commun.*, 361, pp. 794–799.

Wazawa, T.; Ishizuka-Katsura, Y.; Nishikawa, S.; Iwane, A. H. and Aoyama, S. (2006). Grafting of poly(ethylene glycol) onto poly(acrylic acid)-coated glass for a protein-resistant surface, *Anal. Chem.*, 78, pp. 2549–2556.

Williams, S. S.; Hampton, M. J.; Gowrishankar, V.; Ding, I. K. Templeton, J. L.; Samulski, E. T.; DeSimone, J. M. and McGehee, M. D. (2008).

Nanostructured titania–polymer photovoltaic devices made using PFPE-based nanomolding techniques, *Chem. Mater.*, 20(16), pp. 5229–5234.

Wood, R. W. (1902). On a remarkable case of uneven distribution of light in a diffraction grating spectrum, *Phil. Mag.*, 4, pp. 396–408.

Wu, W.; Cui, B.; Sun, X. Y.; Zhang, W.; Zhuang, L.; Kong, L. and Chou, S. Y. (1998). Large area high density quantized magnetic disks fabricated using nanoimprint lithography, *J. Vac. Sci. Technol. B*, 16(6), pp. 3825–3829.

Wulff, G. (1995). Molecular imprinting in cross-linked materials with the aid of molecular templates—A way towards artificial antibodies, *Angew. Chem. Int. Ed.*, 34, pp. 1812–1832.

Xiao, P. F.; He, N. Y.; Liu, Z. C.; He, Q. G.; Sun, X. and Lu, Z. H. (2002). In situ synthesis of oligonucleotide arrays by using soft lithography, *Nanotechnology*, 13, pp. 756–762.

Yamada, S. (2009). Latest trends. In *Plasmonic nanomaterials*, CMC Publisher: Tokyo, Japan.

Yang, C.-Y.; Brooks, E.; Li, Y.; Denny, P.; Ho, C.-M.; Qi, F.; Shi, W.; Wolinsky, L.; Wu, B.; Wong, D. T. and Montemagno, C. D. (2005). Detection of picomolar levels of interleukin-8 in human saliva by SPR, *Lab Chip*, 5, pp. 1017–1023.

Youn, S. W.; Ogiwara, M.; Goto, H.; Takahashi, M. and Maeda, R. (2008). Prototype development of a roller imprint system and its application to large area polymer replication for a microstructured optical device, *J. Mat. Proc. Technol*, 202(1–3), pp. 76–85.

Yu, Y. Y.; Chang, S. S.; Lee, C. L. and Wang, C. R. C. (1997). Gold nanorods: Electrochemical synthesis and optical properties, *J. Phys. Chem. B*, 101, pp. 6661–6664.

Zhang, W. and Chou, S. Y. (2003). Fabrication of 60-nm transistors on 4-in. wafer using nanoimprint at all lithography levels, *Appl. Phys. Lett.*, 83, pp. 1632–1634.

Index